Analog Circuits and Signal Processing

Series Editors

Mohammed Ismail, Khalifa University of Science, Technology
Mohamad Sawan, École Polytechnique de Montréal

More information about this series at http://www.springer.com/series/7381

Ahmed Khattab • Zahra Jeddi • Esmaeil Amini
Magdy Bayoumi

RFID Security

A Lightweight Paradigm

Springer

Ahmed Khattab
EECE Department
Cairo University
Giza, Egypt

Esmaeil Amini
Yahoo Corporation
Santa Clara, CA, USA

Zahra Jeddi
Intel Corporation
Santa Clara, CA, USA

Magdy Bayoumi
The Center for Advanced Computer Studies
University of Louisiana at Lafayette
Lafayette, LA, USA

ISSN 1872-082X ISSN 2197-1854 (electronic)
Analog Circuits and Signal Processing
ISBN 978-3-319-47544-8 ISBN 978-3-319-47545-5 (eBook)
DOI 10.1007/978-3-319-47545-5

Library of Congress Control Number: 2016958309

Printed on acid-free paper

This Springer imprint is published by Springer Nature
The registered company is Springer International Publishing AG
The registered company address is: Gewerbestrasse 11, 6330 Cham, Switzerland

To my great parents, Ahmed Khattab
To my parents and husband, Zahra Jeddi
To my parents and wife, Esmaeil Amini
To my dear students, Magdy Bayoumi

Preface

Radio frequency identification (RFID) is a type of automatic identification systems which has gained popularity in recent years for being fast and reliable in keeping track of the individual objects. In RFID systems, contactless object identification is achieved using radio signals without the need for physical contact as the case with other existing identification technologies such as barcodes. Therefore, a huge number of items can be identified in a short amount of time with high reliability and low cost which makes the RFID technology very attractive for a wide range of applications such as supply chain management, e-health, monitoring humans, pets, animals, and many other objects, toll control, and electrical tagging. Furthermore, RFID technology eliminates the human error and reduces the total cost of the products.

An RFID system typically consists of three main components: A transponder or tag which is implanted on the objects to be identified and stores the objects' identification information such as the object's identification (ID) number, the manufacturer name, and the product type; a transceiver or reader which provides an electromagnetic field in order to activate the tags and read their data through radio frequency waves; and a back-end server which receives and processes the data from readers.

Out of the three main components of RFID systems, tags have the more stringent implementation limitations. In general, there exists three types of tags: passive, semi-passive, and active tags. Active and semi-passive tags are equipped with their own batteries whereas passive tags rely on the radio frequency energy obtained from the reader. Compared to active and semi-passive tags, passive tags have longer lifetime and are smaller and lighter. However, their signal range is shorter than active tags. Passive tag systems are severely constrained in terms of chip area and power consumption as they do not have internal power source. This book focuses on the severely resource-limited passive RFID tags.

Unfortunately, RFID systems face several challenges in their quest to ensure the reliability of the system, quality of service, or reduced system cost. One challenge is the lack of global standardization. As a result of the existing numerous RFID applications, there are many standards for RFID systems. Each standard is designed to fit a specific category of applications. This creates a problem in integrating several RFID systems with each other and makes the manufacturing process harder. Another challenge is maintaining the tag cost as low as possible to contribute in reducing the total cost of the product. However, security is one of the biggest challenges that face any RFID systems. The RFID technology is vulnerable to security attacks by unauthorized reader(s) which can interrogate or modify the information stored in the tags. Due to the limited available resources in RFID tags, providing privacy and security for RFID systems is more challenging than other traditional communication systems. This book is devoted for the security of RFID systems.

RFID security threats are categorized into two main groups: privacy violation attacks and security violation attacks. In privacy violation attacks, the attacker tries to harvest the information stored in the objects by eavesdropping on the communications between the objects and the reader or by tracking them. In security violation attacks, an adversary counterfeits the behaviors of legitimate tags or readers for making undesirable effects such as denial of service. Therefore, it is a necessity to develop mechanisms that provide privacy and security of the communications in RFID systems. This can be achieved via physical privacy protection solution, via authentication, or via cryptography.

Several RFID security physical solutions have been developed such as killing tags, blocking tags, Faraday cages, and active interference. Each of these methods has its pros and cons. For instance, killing a tag will cause the tag to lose its functionality, and hence, it cannot be reactivated. Thus, such a solution considerably reduces the lifetime of tags. Meanwhile, in the blocking tag approach, the attacker cannot have access to tags just in a defined range. Beyond this range, tags are not protected from attacks. In Faraday cage solutions, a wrapper shields the tag from the radio waves which imposes another cost to the system. Unauthorized readers are impeded to have communications with tags in active interference privacy protection solutions. However, sometimes some legal readers get blocked as well in the process. Based on the limitations and disadvantages of the physical security solutions stated above, such methods are only applicable for some specific applications.

Authentication is a process through which an object proves its claimed identity to another communication party by providing some evidence such as what it knows, what it has, or what it is. This process is applicable through only software solutions and it is not possible by physical solutions. In RFID systems, authentication is required in two phases. First, before beginning any communication, both the tag and the reader should verify their identity to make sure that they are contacting with the wished partner. The second phase is when data is exchanged between the two parties to ensure that the exchanged data is intact.

Cryptography solutions keep the communication between two parties private in the presence of third parties. An encryption scheme is composed of five components: a plaintext, an encryption algorithm, a secret key, a ciphertext, and a decryption algorithm. Several encryption solutions have been developed for wireless communication systems to address such security challenges. On one hand, there exist several asymmetric or public key encryption algorithms that use two keys to secure data in networked systems. However, such solutions are not applicable to RFID systems—despite their high security performance—due to the limited processing and power capabilities of the RFID tags. Even existing highly optimized hardware implementation of such algorithms is way beyond what a typical RFID system can afford, such as the hardware implementation of Rabin cryptosystem which offers the best compromises between speed, area, and power consumption. Hence, RFID encryption algorithms must be light enough in terms of area and power to satisfy the resource limitations of RFID systems. Likewise, using hash functions is not suitable for constrained environments since they require significant amounts of resources in their designs, and hence, they are not hardware friendly. On the other hand, several symmetric or private key encryption algorithms have been developed, which are less resource hungry compared to public key encryption algorithms. Even though private key security algorithms promise reasonable security and meet the low resource requirements of RFID systems, they are required to be integrated with other algorithms, such as message authentication code (MAC) algorithms, in order to provide the targeted authentication and integrity services.

In this book, after presenting the RFID security preliminaries, we present the redundant bit security (RBS) lightweight symmetric encryption approach which is suitable for RFID resource-constrained applications. In RBS, the message is intentionally manipulated by distributing redundant bits among plaintext bits, and the location of the redundant bits inside the transmitted data represents the secret key between the sender and the receiver. Meanwhile, there is a relationship between the plaintext data and the redundant data in the RBS algorithm. These redundant bits are generated by a MAC algorithm whose input is the plaintext data. Therefore, these redundant bits can be used for authenticating the message as well. The security level of the RBS approach is adjustable through the number of redundant bits. In other words, there is a dependency between the provided security and the authentication part of the system which distinguishes the RBS algorithm from other existing algorithms. To have flexibility in the number of redundant bits, the implemented MAC algorithm generates variable length outputs. In addition to the number of redundant bits, their values and their positions in the ciphertext are also determining factors in the security of the generated ciphertext. Furthermore, some plaintext bits are also altered based on the value of the encryption key and the redundant bits in order to make the generated ciphertext more secure against attacks. The security of the algorithm is analyzed against existing well-known attacks such as known plaintext, known ciphertext, chosen plaintext, and differential attacks. Experimental and simulation results confirm that the RBS implementation requires less power and area overhead compared to other known symmetric algorithms proposed for RFID systems, especially when the authentication is essential as in harsh operating environments.

RFID Security: A Lightweight Paradigm targets a wide range of readers including but not limited to researchers, industry experts, and graduate students. This book presents the fundamental principles of RFID cryptography that the interested reader will be able to glean information not only to incorporate into his/her own particular RFID security design problem, but also most of all to experience an enjoyable and relatively effortless reading, providing the reader with intellectual stimulation. This book also offers the reader a range of interesting topics portraying the current state of the art in RFID technologies and how it can be integrated with today's Internet of Things (IoT) vision. Readers with theoretical interests will experience an unprecedented treatment of RFID security that takes into account the practical limitations of today's technologies. Meanwhile, readers interested in real-life RFID security implementations will be exposed to a first-of-its-kind lightweight implementation that results in a significant multi-faced performance improvement compared to existing cryptosystems. In simple terms, while several existing RFID cryptography solutions have been developed, they are challenged by the inherent constraints of practical implementation. Analyzing these constraints and proposing an attractive and practical solution to counter these limitations are the basic aims of this book.

Cairo, Egypt — Ahmed Khattab
Santa Clara, CA, USA — Zahra Jeddi
Santa Clara, CA, USA — Esmaeil Amini
Lafayette, LA, USA — Magdy Bayoumi

Acknowledgements

The authors would like to acknowledge Dr. Hong-yi Wu, Dr. Ashok Kumar, and Dr. Mohammad Madani for their precious comments and feedback that helped us further improve the material presented in this book. We also acknowledge Haythem Idriss, Tarek Idriss, Sultan Arabi Sultan, and Shady Elmakhzangy for their help in collecting some material used in the book. The authors would like to warmly thank the Springer editorial team for their support and assistance.

Dr. Khattab would like to dedicate a special expression of gratitude and appreciation to his family, especially his beloved parents, Khattab and Sanaa, for their patience and full emotional support throughout his career. He owes it all to them.

Dr. Jeddi is deeply thankful to her family for their persistent care, trust, and support. She is deeply obliged to her father who taught her to work hard for earning her success and her mother who gave her confidence with her endless kindness and inspiration. She also would like to express special gratitude and thanks toward her husband, Esmaeil Amini, for his kind cooperation and encouragement. She would also like to express her appreciation to Mrs. Masoumeh Abouroshi because she may have never reached this place without her help.

Dr. Amini would like to express special thanks to his wife, Zahra Jeddi, for her encouragement, support, and contribution in preparing this book. He considers it an honor to thank his parents for their thoughtful consideration and unconditional love and support all throughout his life. Finally, he thanks the one above all of us, the omnipresent God, for answering him prayers and giving his the strength to overcome the challenges.

Dr. Bayoumi would like to thank his students, former and current, for enriching his life and keeping him young in heart and spirit. They are making the academic life exciting, interesting, and never boring.

Contents

List of Figures

List of Tables

Part I
RFID Security Preliminaries

Chapter 1
Introduction to RFID

Abstract Radio Frequency IDentification (RFID) is a technology that is being increasingly integrated into many aspects of everyday life. The proliferation of RFID has created a paradigm shift in the way humans, pets, merchandizes, assets, etc., are currently being identified and tracked worldwide. RFID technology utilizes inexpensive wireless RFID chips or tags that store data related to the item. A nearby reader can have access to such stored data. Unlike the related magnetic stripe technology and bar code technology, RFID does not require direct contact nor line of sight contact. This chapter overviews the history and the basics of the RFID technology and its applications.

In the last decade, the desire and need to develop new technologies which support automatic identification procedures for objects and items has grown up rapidly. Such technologies offer enormous productivity benefits such as saving time, reducing error and providing abilities like detecting and tracking. Many modern enterprises and big organization such as Wal-Mart and the United States Department of Defense have made great efforts to improve and apply automated oversight in many applications involved with supporting items tracking, logistics management, supply chain management and access control.

Radio-Frequency IDentification or RFID is one of the automatic identification techniques which identify objects remotely through a radio frequency channel. In fact, RFID is not a very new technology. In the era of World War II, radar was used to "detect" aircrafts when they were still distance away. The problem with radar was the lack of means to identify friendly aircrafts or non-friendly aircrafts. Thus motivated, the Germans noticed that the radio signal reflected back to the base would be different if the pilots rolled their planes while returning to the base. The method that the Germans discovered was actually the first usage of RFID technology—more specifically, the first passive RFID system. Later on, the Identify Friend or Foe (IFF) system was developed by the British. In IFF, every British plane was equipped with a transmitter. When the British planes were returning to the base, they would receive signals from a radar station in the base. After receiving signals from a radar station they transmitted signals back to identify themselves.

Nowadays, thanks to a combination of dropping cost and technology advancement, RFID can be applied in a variety of applications and in new ways. Despite

A. Khattab et al., *RFID Security*, Analog Circuits and Signal Processing,
DOI 10.1007/978-3-319-47545-5_1

the attention gained by RFID systems, privacy issues for users such as clandestine physical tracking of objects and inventorying them are becoming a big concern. Enormous research effort has been done in order to solve this problem. However, most methods request heavy or frequent cryptographic operations on RFID tags, which contradict the low cost demand of RFID tags.

1.1 Automatic Identification

The Automatic Identification or Auto-ID system is a broad term refers to any technology that can identify and locate physical objects automatically by electronically exchanging data and without any human interaction. The goal of using Auto-ID systems is to increase the efficiency and decrease the cost by reducing the required human labor at entering data, and consequently decreasing the number of potential human-caused errors.

Due to the high reliability provided by Auto-ID systems, utilizing them are getting widespread in applications that require tracking items like supply chain and the manufacturing processes from the point of producing the products up to the point where the products are sold or served.

There are various Auto-ID solutions (Fig. 1.1) that are used in industry such as barcodes, chip cards or smart cards, Optical Character Recognition (OCR), voice recognition, biometric (e.g. print screen) and Radio Frequency Identification (RFID) [10]. Selecting the best Auto-ID solution among all of the introduced solutions for particular applications depends on the requirements of the application and also the benefits of the chosen solution. In what follows, each solution will be introduced individually and its strengths and weaknesses are compared to other solutions.

Barcodes are the most common Auto-ID solution in the industry due to their very low cost. A barcode is a small printed image of bars and spaces, attached to items. It is indicating a binary code which identifies the item. To read the data, it is required that the image to be exposed to a scanner. Printing barcodes is easy and cheap which makes the cost of their production very low. Despite the simplicity, universality and low cost advantages of barcodes, they need a direct contact with scanner to be read which makes the speed of reading items low. Also, their readability might be vanished in harsh environment with dirt or moisture.

Smart cards are cards with embedded Integrated Circuit (IC) which is helpful to provide identification, authentication, data storage and processing. Reading data of a smart card is performed through its contact area which makes an electrical connectivity between a reader and the card when the card is inserted into the reader. Smart cards do not have any integrated battery and their required power for communication is provided by the reader. These cards prevent unauthorized reading. However they are vulnerable to harsh environment and they can get affected by dirt. One of the other disadvantages of this solution is the cost of maintaining the readers which is very high [10].

In Optical Character Recognition (OCR), any scanned image of text like handwritten or printed text is converted into digital text and processed. The main

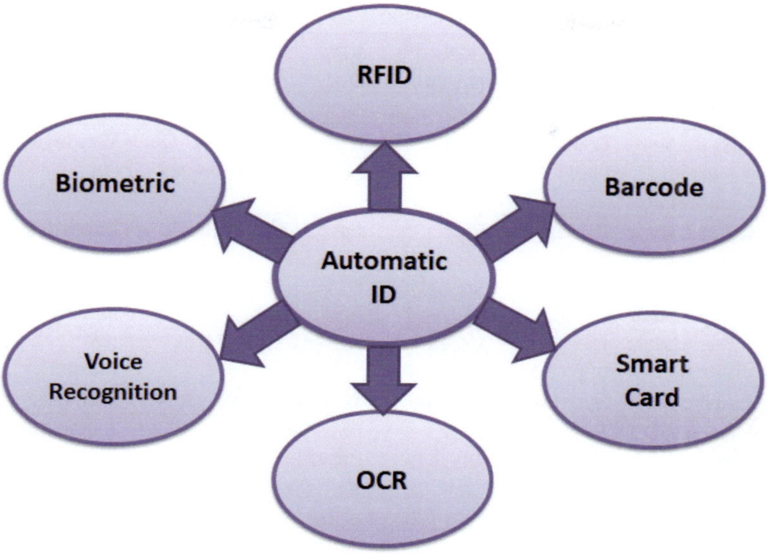

Fig. 1.1 Automatic identification solutions

advantage of this solution lies in handling a high density of information. The most important problem in OCR is the cost of the readers which is high due to their complexity [10].

In voice recognition, the voice of a speaking person is converted into digital data. To recognize the object, this information is compared with the reference patterns recorded before from all objects. This solution works for just humans and utilizing it is not applicable for other objects such as identifying products [10]. The other disadvantage of this solution is the possibility of forgery by using taped voice.

Biometrics are a type of solution by which people are identified by their individual physical attributes such as DNA, finger print, palm image, and facial image. Voice recognition is a subcategory of the biometrics solution with the difference that voice recognition depends only on audio data while other characteristics use image data. In the biometrics solution, direct connection for verifying the identity is required. Similar to voice recognition, this solution is applicable just for humans.

Radio Frequency Identification solution is closely related to smart cards with the main difference that RFIDs can connect to a reader wirelessly when the electromagnetic field is provided by the reader. In this solution, identification is performed using radio signals. Thus, RFID systems do not need physical contact between the reader and the card. This way, a huge number of items can be identified in a short time with high reliability and low cost which makes this method very attractive for applications like supply chain management, e-health, monitoring objects, electrical tagging, etc. RFID tags can be read in a wide variety of circumstances, where barcodes or other optically read technologies are useless. However, this technology with all of its benefits is still costly.

Table 1.1 Comparison of auto-ID solutions [10]

	Barcode	OCR	VR	Biometrics	RFID
Data size (byte)	1–100	1–100	N/A	N/A	16–64 K
Data density	Low	Low	High	High	Very high
Readability by machine	Good	Good	Complex	Complex	Good
Readability by people	Partially	Easy	Easy	Difficult	Impossible
Affected by dirt/ moisture	Strongly	Strongly	N/A	N/A	No influence
Effect of sight distraction	Usage impossible	Usage impossible	N/A	N/A	No influence
Initial costs	Very low	Medium	Very high	Very high	Medium
Unauthorized coping	Easy	Easy	Possible (tape)	Impossible	Impossible
Reading speed	Slow	Slow	Very slow	Very slow	Fast
Max distance reader/carrier	0.5 cm	Under 1 cm	0.5 cm	Direct contact	0.5 m

Table 1.1 presents a comparison of the different Auto-ID solutions based on the different terms. Among all solutions, RFID system gives the best tradeoff compared to all other candidates.

1.2 RFID History and Standardization

The concept of communication using reflected radio energy is quite old and dates back to the origin of the radar technology. The passive communication technology often used in RFID was first presented in Henry Stockman's seminal paper "Communication by Means of Reflected Power" in 1948 [5]. Identify Friend or Foe (IFF) is one of the first applications of radio frequency identification system was developed by the British Royal Air Force during World War II. IFF allowed radar operators and pilots to automatically distinguish friendly aircraft from enemies via RF signals and helped prevent friendly fire incidents.

Electronic Article Surveillance (EAS) was the first commercial RFID application, which was used as a theft prevention system. Such systems were commercially available through companies such as Kongo, Sensormatic and Checkpoint in the late 1960s. These EAS systems typically consisted of a magnetic device embedded in a commercial product and would be deactivated or removed when an item was purchased. The presence of an activated tag passing through an entry portal would trigger an alarm. These types of systems are often used in libraries, music stores, or clothing stores. Unlike RFID, these types of EAS systems do not automatically identify a particular tag; they just detect its presence.

Major progress was picked up in the 1980s and 1990s with varying interests in different parts of the world. Interest in the United States included transportation and personnel access, while European countries were interested in short-range systems for tracking animals, industrial and business applications and electronic toll-collection. The first RFID-based toll-collection system became operational in October 1987 in Alesund, Norway. The increase in the commercial use of RFIDs prompted a need for standards, which led to many standardization activities in the 1990s.

Most of such standards were mainly developed by the International Standards Organization (ISO) and the International Electrotechnical Commission (IEC). ISO, a global organization to which 157 countries belong, develops industry-wide standards in a number of fields. IEC is also a global organization, but it concentrates on standards for electrical, electronics, and related technologies. Initial standardization interests were in animal tracking (ISO-11784 and ISO-11785) and contactless proximity cards (ISO-14443) applications. Enabler in supply chain management spurred a further series of standardization activities. A milestone came in 1996 with the standardization of RFID as a data carrier by the Article Number Association (ANA) and the European Article Numbering (EAN) groups. In 1999, EAN International, and the Uniform Code Council (UCC) of the United States, now both known as GS1, adopted an Ultra-High Frequency (UHF) band for RFID and established the Auto-ID Center at the Massachusetts Institute of Technology. This organization was charged with developing a global RFID standard for product labeling called the Electronic Product Code (EPC) [25].

The Auto-ID Center later evolved into Auto-ID Labs and EPCglobal Inc. The latter is a nonprofit organization, set up by UCC and EAN International, pursuing the commercialization of EPC technology. The recent advances in silicon technology made RFID tags cheap and reliable. Thus, the first decades of the twenty-first century see the world moving toward the technology's widespread and large-scale adoption. A major landmark was the announcement made by Wal-Mart Inc., in the United States, to mandate RFID for its suppliers in "the near future," at the Retail Systems Conference held in Chicago in June 2003. This was followed by the release of the first EPCglobal standard in January 2005. Up to the date of writing this book, more than 1000 Wal-Mart locations have already implemented EPC RFID standard [7, 25].

Due to the wide spectrum of RFID applications and use cases, several standards have been developed for such RFID applications. Table 1.2 summarizes the existing RFID standards and their targeted application.

1.3 RFID Applications

RFID tags bring huge benefits over many systems since they have the ability to be read if they pass near a reader even if it is covered by objects or not visible like when it is in a container or a box. Also, hundreds of tags can be read at the

Table 1.2 RFID standards

Standard	Operating frequency	Applications
ISO 14223	135 kHz	Animal identification
ISO 11785	135 kHz	Animal identification
ISO 11784	135 kHz	Animal identification
ISO 18000-2	135 KHz	Item management
ISO 18092	13.56 MHz	Near-field communications (NFC)
EPC HF 1	13.56 MHz	Electronic product code (EPC)
ISO 15693	13.56 MHz	Contactless chip cards
ISO 14443	13.56 MHz	Contactless chip cards
ISO 18000-3	13.56 MHz	Item management
ISO 18000-7	433 MHz	Item management (active tags)
EPC UHF 1	900 MHz	Electronic product code (EPC)
EPC UHF 0	900 MHz	Electronic product code (EPC)
ISO 18000-6	900 MHz	Item management
ISO 18000-4	2.4 GHz	Item management

same time instant. These advantages offer new solutions to a variety of applications. Analyzing the RFID market in many different ways, technical experts expect that retail dominates the market in the near future (Fig. 1.2) [15]. Thus, retail companies are required to move towards RFID system to avoid losing their profits. The suppliers of other sectors in this pie will receive the benefits of RFID by providing a secure and enduring support for their customers, considering anti-counterfeiting RFID for drugs, error-preventing RFID on hospital instruments and anti-terrorism measures in airports.

In a world where everyday objects carried RFID tags, remarkable things would be possible. In this section, we briefly discuss a few possibilities (among the myriad that the reader might dream up). The most important RFID applications focus on logistics, supply chain management, toll system, tracking, ticketing, health care, security, and identification systems [27].

1.3.1 Logistics and Supply Chain Management

One of the most famous RFID applications is supply chain management. By attaching RFID tags to each product, tool, resource and item, all items are tracked by RFID readers from the manufacture to the points of sales. Hence, the manufacturers will be able to get better demand signals from customers. RFID simply offers the potential to improve product life-cycle management, and quality control with the aim of helping retailers to provide the right product at the right place at the right time and consequently to maximize sales and profits. Megatrux, a top 100

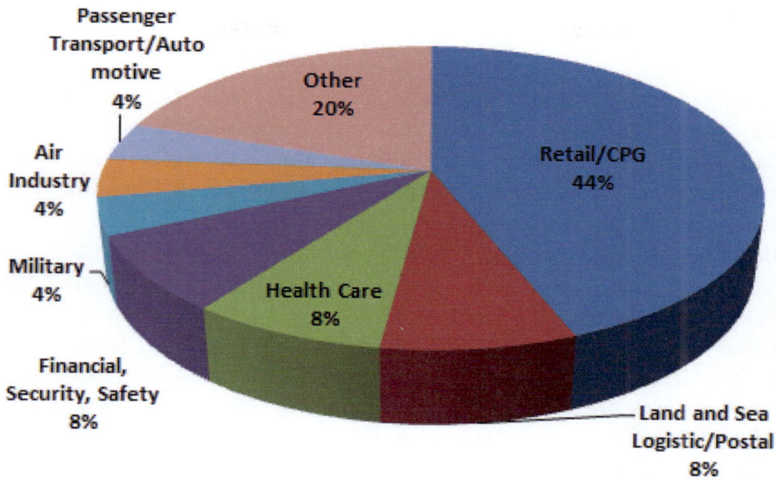

Fig. 1.2 Global RFID market value in 2016 as reported in [15]

logistics company in the world, has applied Motorola RFID plan to its supply chain management [21]. Their customer services have been improved while the costs have been significantly reduced.

RFID also has a great value in product delivery. We could track the handing process and current location of the product from pick up to delivery with an RFID tag attached on it. It helps to relieve incorrect delivery owing to human mistakes. RFID does not require one to one line of sight reading which is required for barcode. This reduces the time and cost of reading tags for a large batch of goods. DHL started developing a global Information Technology (IT) infrastructure for tracking all packages with RFID tags attached worldwide by 2015 [3]. The Australia Post begun processing RFID-tagged envelops in domestic mail service in 2005 [8]. Federal Express runs a pilot test on the application of RFID technology to track packages' temperature, location, humidity levels and delivery status [9].

In apparel industry, RFID is applied in manufacturing, distribution and retail. An RFID system provides inventory visibility and enhances customer shopping experience. The RFID system is able to know which clothes being picked up by customers most frequently and to provide customized advertisement on picked clothes according to the RFID tags. American Apparel applied RFID system to eight of their stores. It helps to save about 60–80 h per week in labor and reduce products out-of-stock owing to unawareness [9]. Japanese apparel manufacturer Sankei utilizes a RFID system in clothing manufacturing to track clothes during the manufacturing process and facilitate online sales [1].

1.3.2 Ticketing

Owing to the small size and flexible antenna of RFID tags, they have been widely applied in e-tickets for exhibitions, stadiums, theme-parks and entertainments. Compared with ordinary ticket, the e-ticket is more resistant to fake tickets and facilitates contactless automatic identification. Moreover, it also provides extra functions, such as guess allocation, flow controlling of people, etc. There is no fake ticket being found in World Cup 2006 because RFID tags were applied in its tickets. It attracts the exhibition sponsors' attentions. Beijing 2008 Olympic Games increases the application of RFID tags in its tickets [13]. Personal information is embedded in tickets for opening and closing ceremonies of the Beijing Olympic Games. The ticket holder's photograph, passport details, home addresses, e-mail and telephone numbers are stored in her/his ticket [20]. RFID technology is a secure, reliable and convenient tool for personalized information services. For Beijing Olympic Game staffs and players, the RFID tagged ticket helps in registration on arrival, security identification and payment.

In the light of the previous successful applications in world class games, RFID cards was adopted in Shanghai World Expo [23]. The total sales of tickets will surpass 620 million pieces. It was a RFID project with the largest amount of tags being used, the highest quality requirements, and the longest time span in the world.

RFID enjoys advantages in speed, accuracy and convenient over traditional tickets. Therefore, RFID tagged e-tickets will gradually replace traditional tickets and facilitate intelligent applications in exhibitions, games and theme-parks.

1.3.3 Health Care

Health care demands for extreme accuracy in drug distribution, handling and processing. Institute of Medicine (IOM) reported that the human carelessness is one of the major causes of medical errors [16]. RFID technology would aid the medical staff in performing their duties and reduce medical errors [5]. Examples include automating the admission, screening and treatment processes, and enhancing communications between caregivers and support teams [14]. Another major RFID application in health care centers is the access control of staffs and patients. Each of them is issued an RFID card recording their access permissions. The control center can locate patients or staff members by the readers deployed in different locations. The center can also track patients and control the access of medical equipments and restricted zones. Some hospitals tag all equipments and use the tags to track equipments. This helps managing inventory and ensuring proper maintenance of equipments. Some hospitals also use RFID tags on new-born babies to ensure their identification. If someone attempts to take the baby away from the hospital without authorization, the system will alert the hospital staff [26]. In drug managements, RFID technology could be used to identify fake drug and monitor real-time stock.

In addition, illegal dumping medical waste can be punished by tracking RFID tags in the medical waste.

In hospitals, RFID also works with other sensors to collect patients' health information. RFID technology offers a great market potential in this area. A report from ID TechEx showed that the market value in the American health care industry reached $ 86.3 billion by 2010 [19].

1.3.4 Security and Identification

Security and personal identification applications, in which RFID tags are embedded in ID cards, is another major application of RFID [24]. Now, a new generation of ID cards and student cards adopt RFID technology. The USA passport is embedded with a RFID tag inside. The RFID tags provide a more reliable storage of identification information compared to magnetic strips. Moreover, many organizations use RFID cards to control different levels of access according to different security levels granted to the card owner. Readers are deployed at the building entrance and only allow authorized person to get access. Furthermore, some high-security applications allow identifying people by injected RFID chips under their skin for use in a variety of settings, including financial and transportation security, military and government security to control accesses to secure areas. However, injecting RFID under human skin raises several ethical questions which renders its wide spread.

1.3.5 Toll Systems and Payment Applications

Toll systems using RFID technology to facilitate electronic toll collection is widely deployed, especially in highways and car parks. The RFID toll system enables vehicles to check-in and check-out automatically under a fast, contactless, secure and convenient environment. However, cars must be queuing up and pass through the toll system gate one by one [2]. Nonetheless, RFID based automatic toll systems relieve the traffic jam problem caused by the long queue in human manned toll station.

Other RFID payment applications are currently being widely adopted as a convenient way for payment such as contact-less credit cards. However such RFID payment applications requires high levels of security.

1.3.6 Tacking Applications

Several tracking applications exploit the advantages of the RFID technology such as:

- **Asset tracking:** the location of tagged assets like health care facilities or a laptop can be instantly determined anywhere within the help of RFID technology. This application is also very useful in some services like postal services, and monitoring vehicle traffic.
- **Animal tracking:** this application keeps the track of livestock to help prevent disease outbreaks. It also can be used by pet owners to keep track of their animals when they are lost.
- **People tracking:** this application is required in hospitals and jails. In a hospital, this technology can help to track special patients who need special or mental care and also for new born babies.

1.3.7 RIDF and Smart Objects

There are some trending applications which are becoming applicable because of the RFID technology that can be referred to as smart objects. For example a smart oven which knows how to cook pre-packaged food by reading the cooking instructions stored on the RFID tag of the food. Other example applications that take advantage of RFID technology include—but not limited to—smart products, smart appliances, RFID-enabled mobile phones and recycling plastics [17]:

- **Smart Products:** Clothing applications, CDs, etc. tagged for store returns.
- **Smart Appliances:** Refrigerators that automatically create shopping lists. Also, closets that tell you what clothes you have available, and search the Web for advice on current styles, etc. And, one such application is VistaCrafts RFIQ available in Japan, which comes with 24 recipe cards. The pan reads the card you show and "tells" the cook top what to do to perfectly monitor each cooking step and perfectly reproduce the most difficult recipes. Each pan handle is embedded with an RFID chip that uses a proprietary signal to communicate with coordinated chips in the cook top and special recipe cards that monitor each cooking step for a particular dish.
- **RFID-Enabled Mobile Phones:** Scan a movie poster to learn show times, scan consumer product to get price quotes, etc.
- **Recycling plastics that sort themselves.**

1.4 RFID System Overview

In general, each RFID system consists of three parts (Fig. 1.3): (1) a transponder or tag that carries the ID data, (2) a transceiver or reader to interrogate the tag and extract information from it, and (3) a back-end server with a software application acting as an interface between the user and the RFID system.

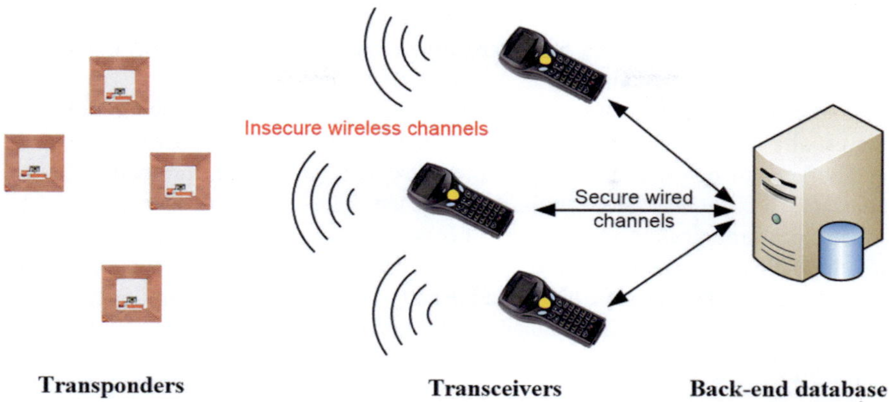

Fig. 1.3 RFID system architecture

Fig. 1.4 Generic block diagram of RFID tags

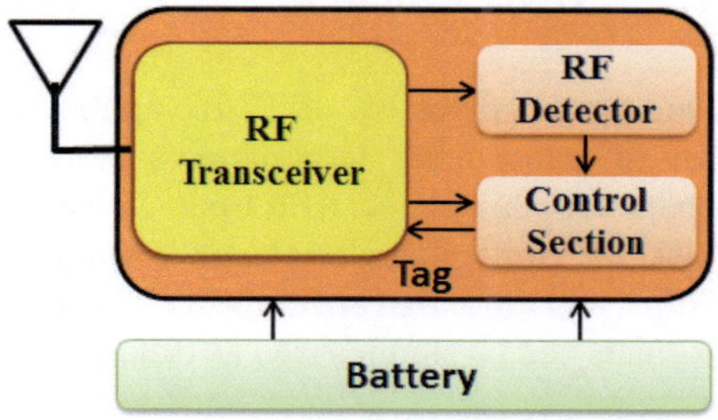

An RFID transponder or tag is a data carrying device that is added to items to be later interrogated by an RFID reader. The main purpose of a transponder is to carry the identification information of the object it is attached to. The tag is attached to or embedded in an object to provide unique identification for it. It contains some information associated with the corresponding object. This information can be either as short as few bits or be a collection of data such as the identity code for animals, the expiration date for groceries and the personal medical information for people.

A tag can be as simple as being composed of an electronic circuit with unique identifiers and one antenna, used for communication. Such tags are called passive (i.e., battery-less). As shown in Fig. 1.4, more avalanched tags can be composed of an antenna, a radio frequency (RF) transceiver section, an analog detection and/or rectification section which detects, and in passive tags, rectifies the RF power into an equivalent DC voltage, and digital control section that is either a microprocessor or some other digital system. In addition, a tag can use a battery in the case of active and semi-passive tags.

A transceiver or reader is a two-way radio transmitter-receiver that both receives and transmits radio waves unlike tags which transmit signals only in response to received signals. The reader has a powerful antenna and a power supply, surrounding itself with an electromagnetic field in order to activate tags and read their data through radio frequency waves. The collected data from tags by the reader is sent to the back-end server. This server contains a database of tags' information. The received data are stored and processed in the back-end server.

The channels between the reader and the back-end database are wired links that are usually assumed to be secure. On the other hand, both the reader and the back-end server are powerful enough to apply strong cryptographic protocols. On the contrary, the channels between the tags and the reader are wireless channels. The wireless communication is in danger of eavesdropping by adversaries which make it vulnerable to a variety of attacks. Handling contemporary cryptographic protocols in RFID tags is not possible since they usually have restricted capabilities in every aspect of computation, communication and storage because of their extremely low production cost.

1.5 RFID Construction Formats

Due to the diversity of RFID applications, each RFID system has a different set of transponder requirements that put different constraints on the physical characteristics of the RFID tag. Some common tag construction formats include disks or coins, glass or plastic housing, keys and key fobs, smart labels, coil-on-chips, and those that are embedded in smart cards [10, 11]. The various construction formats can be summarized as follows.

- **RFID Disk and Coins:** This is the most common RFID tag construction format as shown in Fig. 1.5. Epoxy resin molding can make this format withstand higher temperature levels.
- **RFID with Glass/Plastic Housing:** Some applications, such as animal tracking and identification, require injecting the RFID tag underneath the animal skin. Hence, RFIDs developed for such applications are typically enclosed in a glass

Fig. 1.5 An RFID coin

Fig. 1.6 Injectable RFID within a glass housing

Fig. 1.7 Typical RFID key fobs used in secure areas

Fig. 1.8 Smart labels has the RFID transponder printed under the print-coded label

or plastic capsule as shown in Fig. 1.6 such that it can be injected underneath the animal skin with no harm to the animal nor to the tag itself.

- **RFID Key Fobs:** Such an RFID construction format shown in Fig. 1.7 is widely used for immobilizers or door locking applications for high security areas.
- **RFID Smart Labels:** This format is a paper-thin transponder under a conventional print-coded label as shown in Fig. 1.8. The labels are typically made of paper, fabric or plastic. In this format, the tag is produced by either printing or etching.

Fig. 1.9 A typical RFID smart card

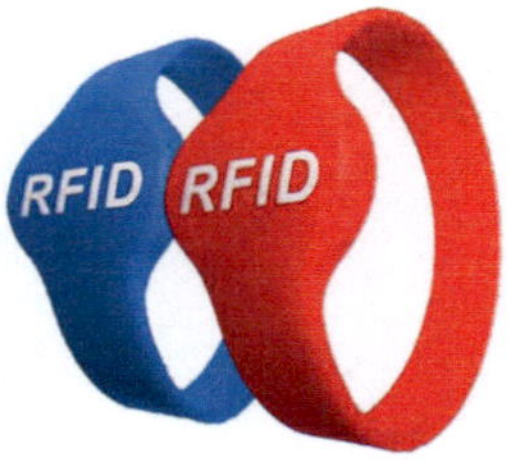

Fig. 1.10 A typical RFID wristband

- **RFID Smart Cards:** Contact-less smart cards have several applications ranging from contact-less access cards to contact-less credit cards. Such cards facilitate the communication transaction without wiping a magnetic stripe. Hence, an embedded chip and a simple antenna are built inside the card to realize such RFID systems as shown in Fig. 1.9.
- **RFID Wristbands:** This is another RFID construction format (depicted in Fig. 1.10) that is widely used for identification. The RFID transponder is typically embedded in a durable and waterproof material.

1.6 RFID Classifications

RFID tags can be classified to many types according to the power source, memory, radio frequency ranges and the way they communicate with the reader.

1.6.1 Communication Mechanism

Based on communication mechanism between the reader and tags, RFID systems are classified to two types [6, 18, 22]:

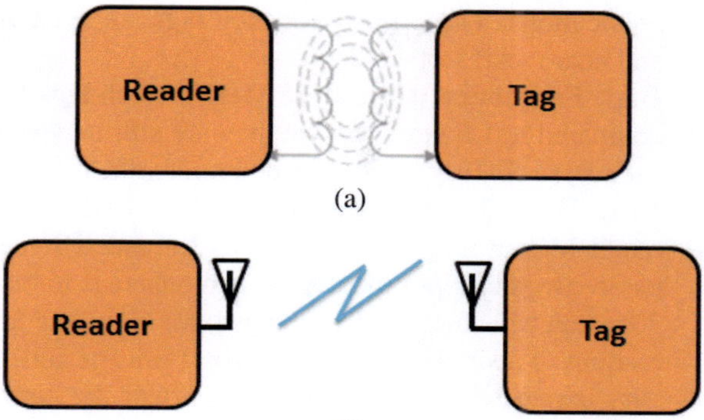

(a)

(b)

Fig. 1.11 Possible communication approaches used in of RFID systems. (**a**) Communication by induction. (**b**) Communication by propagation

- Induction or Near-Field Communication: The reader reads the data stored in the RDIF tag using inductive coupling as shown in Fig. 1.11a. This necessitates that the reader be in a close proximity of the tags.
- Propagation or Far-Field Communication: The reader communicates with the tags by propagating electro-magnetic waves as shown in Fig. 1.11b. Therefore, the reader can communicate with tags that are farther away compared to induction-based tags. However, the complexity and the hardware requirement of such systems are higher as they employ transceiver chains that require power sources.

1.6.2 Memory

Based on the memory, RFID tags can be categorized into two main categories [6, 18, 22]:

- Tags with read only memory: These tags allow only read operations to retrieve the stored data.
- Tags with read/write memory: These tags allow both read and write operations. Hence, the stored data can be changed if needed unlike the read only tags.

1.6.3 Operating Frequency

Based on the operating radio frequency range, existing RFID tags typically operate in four frequency ranges [6, 18, 22]:

- **Low Frequency (LF, 30–500 kHZ):** The communication ranges of such tags are approximately half a meter and are mostly used for short reading range applications. These low frequency tags are least affected when applied on wet and near metal surfaces.

- **High Frequency (HF, 10–15 MHZ):** Such tags have higher data transfer rates compared to LF tags, and yet they are still inexpensive. They are typically used for access control, items or product identification, etc.
- **Ultra-High Frequency (UHF, 850–950 MHZ):** UHF tags have significantly much higher ranges compared to LF and HF tags. Their typical ranges for passive tags is approximately 3–6 m, whereas for active tags ranges of more than 30 m can be achieved. These tags have high data transfer rates which enable the reading of a single tag in a very short time period. These tags are comparatively very expensive. Fluids and metals affect the performance of these tags. UHF frequencies can be different for different countries and require permits.
- **Microwave (μW, 2.4–2.5 GHZ and 5.8 GHZ):** The microwave reader rate is high and even higher than UHF tags. At such microwave frequencies, the reading rates are not the same on wet areas and near metals. These frequencies offers better results in applications such as vehicle tracking, within a tag's reading range of 1 m.

Table 1.3 compares the different RFID systems based on their operating frequencies. Recall that, as the operating frequency decreases, the communication range decreases. Furthermore, the reduction in the operating frequency increases the antenna length, and hence, the size of tag. Hence, LF tags have the smallest rates, cheapest price, and lowest coverage relative to HF tags, UHF, and microwave tags, respectively. On the other hand, LF tags work properly in the presence of fluids and metals compared to HF tags. Microwave tags have the highest transmission rates, smallest reading time (thus, it is proper for tagged objects which high mobility speeds), and highest coverage area.

Table 1.3 Tag frequencies and reading distances

Band	Frequency	Distance	Energy Transfer
Low frequency (LF)	125 kHz	1–90 cm, typically around 45 cm	Inductive coupling
High frequency (HF)	13.56 MHz	1–75 cm, typically around 40 cm	Inductive coupling
Ultra high frequency (UHF)	865–868 MHz 902–928 MHz 433 MHz	Up to 9 m	Electromagnetic coupling
Microwave (μW)	2.45–5.8 GHz	Typically 0.3–0.9 m	Electromagnetic coupling

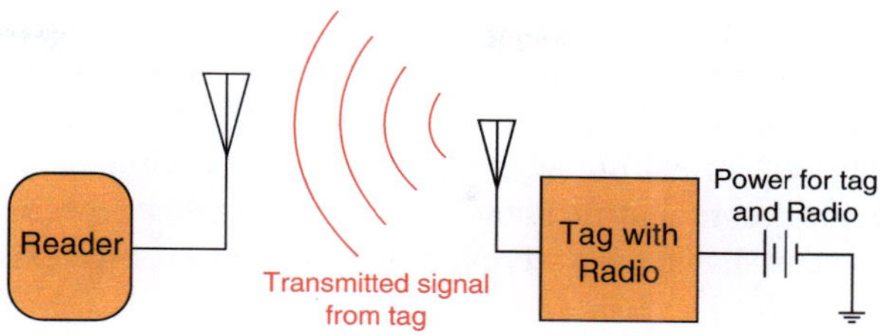

(a)

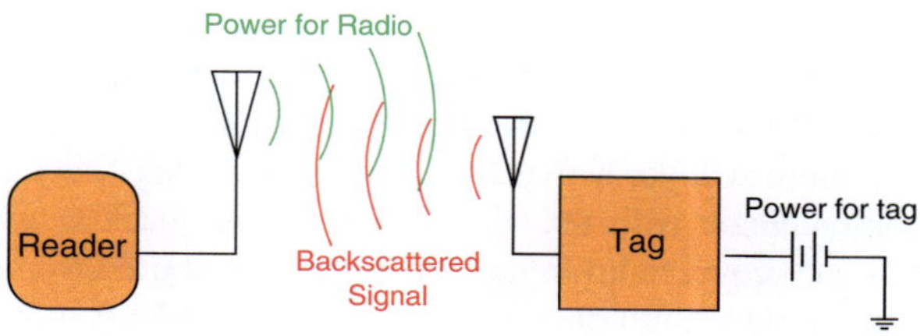

(b)

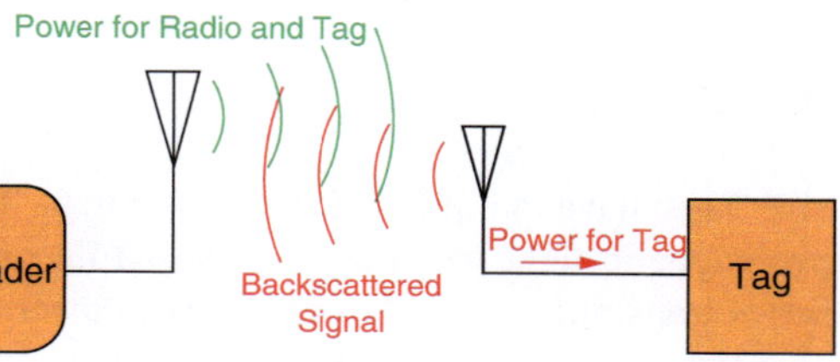

(c)

Fig. 1.12 Different types of RFID tags. (**a**) Active tag. (**b**) Semi-passive tag. (**c**) Passive tag

1.6.4 Power Source

RFID tags are classified according to their embedded power source to three categories: active tags, semi-passive tags and passive tags as shown in Fig. 1.12.

1.6.4.1 Active Tags

In active tags, a radio signal transceiver is embedded along with a power source, usually in the form of a small battery to power it (Fig. 1.12a). Because of the on-board battery, active RFID tags can initiate communication and activate themselves regardless of the presence of a reader in their vicinity. However, active tags usually remain in a low power state until they detect the presence of an RF field being sent by a reader in order to conserve the battery. Whenever the tag leaves the vicinity of a reader, it returns back to the low power state again.

Thanks to the equipped battery, active tags can cover longer ranges compared to other type of tags. Therefore, these tags can be read by the reader while they are much farther away. However, their lifetime is restricted by the capacity of their battery. Even though some of them are built to have up to few years life span, they still have limited lifetimes. Due to these characteristics, active tags are usually utilized in real time systems to measure environmental parameters like humidity, temperature and pressure. Compared to other types of tags, active tags are more expensive and have more limitations because of the existence of the battery.

1.6.4.2 Semi-Passive Tags

Semi-passive tags have their own power supply that supports the integrated microchip only. When the battery is discharged, these tags cannot transmit signals any more. Unlike active tags, semi-passive tags have no active transmitter and to communicate with the reader they use the backscatter technique (Fig. 1.12b). In this technique, radio frequency energy transferred from the reader are gathered and altered to transmit data in a way that the reader can detect. Therefore, they cannot initiate communication.

1.6.4.3 Passive Tags

Passive tags have no internal power source. They draw their power from the electromagnetic field generated by the RFID reader (Fig. 1.12c). They have also no active transmitter and rely only on the power that comes from a reader's signal. Passive tags are inactive unless a reader activates them. Compared to other types of tags, passive tags are cheaper and smaller while the covered range is shorter. Since passive tags do not require having any battery to support their computation and communication, they can stay usable for very long periods of time. Due to these features that make them suitable for a wide range of applications, passive tags are the most common type of tags in the market (Fig. 1.13). Moreover, passive tags can tolerate environmental conditions while these conditions limit the use of tags with on-board batteries. However, in passive tags, the power required for computation and communication is limited by the obtained power from the field. Some solutions have been given to increase the obtained power in the tags. One of such solutions is increasing the antenna gain of the tags which helps to gather more energy from the field. Because of having a limitation on the size of the tag, this solution is impractical. Increasing the power of the field is another solution. However, the maximum strength of the sent signals by readers is limited by law. Due to the nature of RFID tags, designers confront many technical limitations to deal with such as:

- Limited power consumption
- Limited area
- Limited execution time
- Limited backward channel
- Limited memory access

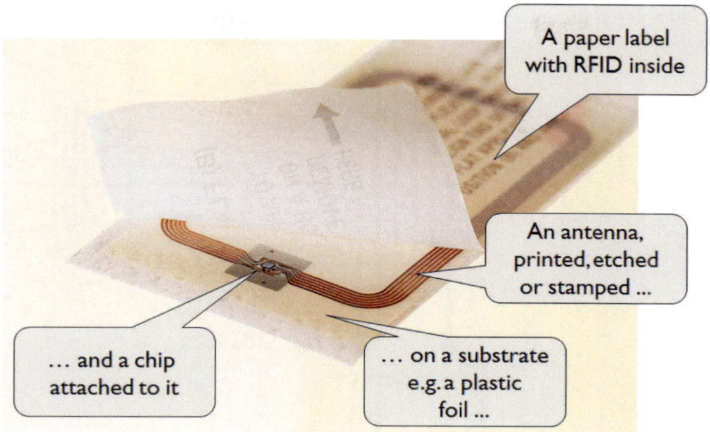

Fig. 1.13 A typical passive tag

Table 1.4 Tag frequencies and reading distances

Tag type	Passive	Semi-passive	Active
Power source	Incident energy	Battery	Battery
Communication type	Response to reader	Response to reader	Initiation/response to reader
Maximum range	10 m	>100 m	>100
Memory	Read only	Read only	Read-write
Relative cost	Least expensive	More expensive	Most expensive
Example application	EPC	Electronic tolls	Large-asset tracking

Table 1.4 summarizes the main difference between passive, semi-passive and active tags. Due to the widespread of passive tags and their unique battery-less operation feature, the next section further explains their operation mechanism.

1.7 How Passive RFID Tags Work

The communication between a passive tag and a reader takes place through transferring energy and data. Energy, provided by the reader, is transferred to the tag using coupling via electromagnetic fields [12]. To receive energy, RFID tags can use both the electric field and the magnetic field or one of them. Passive RFID tags do not have any energy for communication until they enter one of these fields. As soon as tags pass through the field, they are able to draw enough power from the field to become activated.

Based on the provided field, there are different methods for transferring data from the tag to the reader. One of the contemporary techniques is backscattering which was described before. In this method, the reader transmits a continuous wave of radio frequency signal into the environment. When a tag enters in this area, it

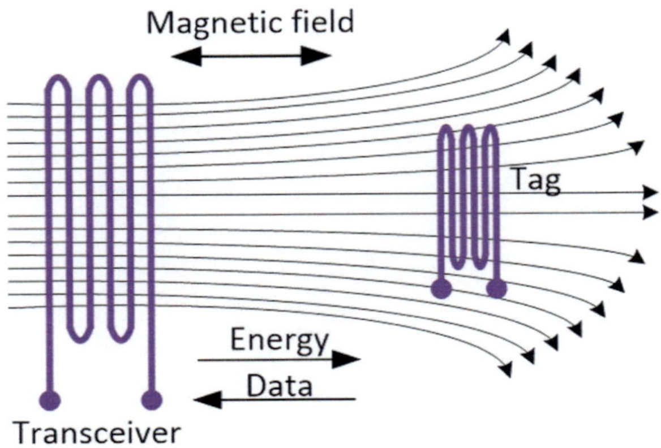

Fig. 1.14 Inductive coupling

receives the reader's signal and demodulates it. The transmitted wave consists of commands to inform the tag what operations to perform. In reply, the tag modulates its response and sends it back to the reader.

Inductive coupling is another common method for transferring energy to passive tags (Fig. 1.14). This method is based on the fact that when a conductor appears in a magnetic field, the magnetic field produces a current flow in the conductor [4]. In this method, the antenna of the reader provides the magnetic field and the tag plays as a conductor. When the tag enters the magnetic field, its antenna generates a current into the tag to power it up. Magnetic fields are utilized in low frequency (LF) and high frequency (HF) RFID tags while the distance between the tag and the reader is short.

The electromagnetic coupling method is similar to the inductive coupling method with the difference that instead of using a magnetic field, an electromagnetic field is utilized which covers a longer distance for transferring energy to tags. Ultra high frequency (UHF) and microwave tags use this method. Table 1.3 summarizes the used methods for energy transferring in RFID tags based on their operating frequency.

1.8 RFID Systems Advantages and Challenges

RFID technology has gained significant widespread over the years due to its multifaced advantages. Yet, the RFID technology is facing numerous challenges. This section is devoted for the advantages and challenges of RFID systems.

1.8.1 Advantages of RFID Systems

RFID systems are going to replace barcode systems and other traditional identification systems [18]. The following points summarizes the main advantages of RFID systems that support such a claim:

- RFID system does not necessitate the involvement of humans in the identification process. This reduces the number of employees, and consequently, eliminates the human error and reduces the total cost.
- RFID system can operate even in the absence of line-of-sight communication between the tags and the reader. Hence, the RFID tag placement have less restrictions compared to barcode systems and other automatic identification systems.
- RFID readers are capable of simultaneously reading multiple tags.
- RFID systems have much longer read ranges relative to barcode systems and other traditional identification systems.
- RFID systems are more reliable than the traditional identification systems such as barcode systems.
- Unlike traditional automatic identification systems, RFID tags have the capability of storing additional information besides the tag ID.
- RFID systems open the door for adding sensing capability to the tag to sense the surrounding environment conditions (e.g., temperature, humidity, etc.) and storing the sensed inform in the tags. Such a capability does not exist neither in barcode systems nor other automatic identification systems.

1.8.2 Challenges to RFID Systems

However, RFID systems face several challenges to ensure the reliability of the system, quality of service, or system cost. The most prominent RFID system challenges are [18]:

- **Standardization:** As a result of the existence of many different RFID applications, there emerged many standards to regulate the implementation of such RFID systems. Each standard is specifically designed to fit a specific category of applications. This creates a problem in integrating and inter-operating such heterogeneous RFID systems and makes the manufacturing process harder.
- **Component Cost:** One of the advantage of RFID systems is that they contribute in reducing the overall cost of the system. However, this poses a challenge in the cost of the RFID tags and readers themselves. For example, RFID tag cost should be in the order of only few US cents.
- **Collision:** One of the benefits of RFID systems is that readers can read several tags at the same time. Consequently, the packets of the different tags can collide with other tags' packets. Thus, the readers have to apply anti-collision techniques

to resolve such collisions and to decrease the system latency. There are two main anti-collision protocols categories: ALOHA (either Pure ALOHA, slotted ALOHA, or framed slotted ALOHA) protocols, and tree-based protocols such as Tree Splitting, Query Tree, Binary Search, Bitwise Arbitration [18].

- **System Security:** One of the biggest challenges that faces any RF system is its security. Since RFID systems use wireless means of communication between the reader and tags, the RFID systems may be faced with eavesdropped, counterfeiting, playback and tracking threats, bringing up communications security issues, especially privacy leak. Due to the importance of securing RFID system, this book focus on such a topic. More specifically, Chaps. 2 and 3 mainly cover the different RFID security threats and their existing solutions, respectively.

1.9 Book Organization

RFID Security: A Lightweight Paradigm consists of two parts. In Part I: RFID Security Preliminaries, the basics of the RFID technology, is briefly introduced in Chap. 1 along with describing its components, types of transponders and their limitations in performing communications. Chapter 1 also overviews the numerous application domains of the RFID technology. It also describes the main advantages and challenges of such a technology.

In Chap. 2, the several security attacks threatening RFID systems and their potential physical solutions are presented. Since this kind of solutions is not able to provide security and privacy for their consumers, software solution called cryptography is recommended in order to solve this problem by preventing attackers from having access to the tags' data without having the secret key. This solution also has the advantage that providing other services such as integrity and authentication will be feasible.

Chapter 3 presents the basic concepts of cryptography with emphasis on lightweight cryptography designated for resource-constrained designs such as RFID systems. A survey of the existing cryptosystems which are compatible with this definition is presented. For each of the existing cryptosystems, the performance under the possible attacks to which they are vulnerable is investigated. At the end of the chapter, the results of their hardware implementation on different platforms are given.

Part II of the book entitled Lightweight RFID Redundant Bit Security is dedicated to the Redundant Bit Security (RBS) lightweight approach developed for the Internet of Things (IoT) applications. The RBS algorithm is based on inserting redundant bits is introduced in Chap. 4. The level of provided security of the RBS algorithm, the location of redundant bits, the value of redundant bits and the method of appearing plaintext in the ciphertext are defined. This is followed by the hardware implementation of the RBS cipher. The RBS hardware implementation consists of two parts. The first part implements a redundant bit generator which is adapted from a Message Authentication Code (MAC) generator. Since the original MAC is designed for stream ciphers, it has to be modified to make it compatible with block ciphers. The second part is implementing encryption/decryption ciphers. This part of the hardware implementation is integrated with transmission and reception parts of an RFID transponder.

Chapter 5 describes the powerful and common security attacks such as known-plaintext, chosen plaintext, related key attacks, etc. Then, the chapter is devoted for illustrating how the RBS algorithm is resilient against these kinds of attacks.

In Chap. 6, the results of the RBS hardware implementation is presented and its one-dimensional and multi-dimensional performance metrics in ASIC design such as area, power consumption, energy and hardware efficiency are evaluated. Afterwards, these results are compared with other existing lightweight cryptosystems discussed in Chap. 3. Since RBS cipher provides authentication for all of messages, this comparison is performed in two categories. First when none of competitor ciphers support the authentication service and second, when all of them do.

Chapter 7 discusses how to integrate the lightweight RFID technology with IoT systems and the pros and cons of such integration from the security point of view. This chapter explains how does the unique characteristics of our RBS lightweight cryptosystem makes it a strong candidate for RFID security in the IoT applications.

References

1. Apparel update-January 2009. RFID Monthly. http://www.rfid-monthly.com/?tag=apparel (2009)
2. Apriso Corporation: Japanese sankei implements apriso's FlexNet. fire2fashion. http://www.fibre2fashion.com/news/textiles-technology-news/newsdetails.aspx?news_id=52554 (2008)
3. Bacheldor, B.: U.N.'s universal postal union gears up for large RFID pilot. RFID J. http://www.rfidjournal.com/article/print/4504 (2008)
4. Brown, M., Zeisel, E., Sabella, R.: RFID+ Exam Cram. Que, Indianapolis, IN (2006)
5. Cangialosi, A., Monaly, J., S.C., Yang: Leveraging RFID in hospitals: patient life cycle and mobility perspectives. IEEE Commun. Mag. **45**(9), 18–23 (2007)
6. Chauhan, M., Sharma, E.: A survey on RFID technology. Int. J. Res. **1**(10), 1316–1322 (2014)
7. Chawlaand, V., Ha, D.S.: An overview of passive RFID. Int. J. Comput. Electr. Eng. **45**(9), 11–17 (2007)
8. Collins, J.: Aussies track mail service via RFID. RFID J. http://www.rfidjournal.com/article/view/2014/1/1 (2014)
9. Dignan, L.: FedEx couples google earth with active package tracking. ZDNet (2007). http://www.fedexaminer.com/FedEx/modules.php?name=News&file=article&sid=172
10. Finkenzeller, K.: RFID Handbook. Wiley, West Sussex (2003)
11. Grover, A., Berghel, H.: A survey of RFID deployment and security issues. J. Inf. Process. Syst. **7**(4), 561–580 (2011)
12. Glover, B., Bhatt, H.: RFID Essentials. O'Reilly, Sebastopol (2006)
13. Guangjin, L.: RFID application in 2008 Olympic Beijing. Radio Freq. Identif. Technol. Appl. **29**(4) (2008)
14. Harrop, P., Das, R.: RFID forecasts, players and opportunities 2005–2015. ID TechEX (2005)
15. IDTechEx. http://www.idtechex.com (2016)
16. Institute of Medicine: Crossing the Quality Chasm: A New Health System for the 21st Century. Institute of Medicine Publication/National Academy Press, Washington (2001)

17. Kannouf, N., Douzi, Y., Benabdellah, M., Azizi, A.: Security on RFID technology. In: Proceedings of the International Conference on Cloud Computing Technologies and Applications (2015)
18. Kaur, M., Sandhu, M., Mohan, N., Sandhu, P.S.: RFID technology principles, advantages, limitations & its applications. Int. J. Comput. Electr. Eng. **3**, 151–157 (2011)
19. Koh, R., Schuster, E., Chackrabarti, I., Bellman, A.: Securing the pharmaceutical supply chain. Auto-ID Center, Mit-AutoID-WH-021 (2003)
20. Lee, J.: First RFID lap counters, now microchipped olympic tickets? SpeedEndurance. http://speedendurance.com/2008/05/31/first-rfid-lap-counters-now-microchipped-olympic-tickets/ (2008)
21. Motorola: The next-generation warehouse megatrux improves service and reduces costs with RFID. RFID World, Rancho Cucamonga. http://www.bendercomm.com/dealer-downloads/CS_Megatrux_1007.pdf
22. Qing, X., Goh, C.K., Chen, Z.N.: Segmented loop antenna for UHF near-field RFID applications. Electron. Lett. **45**(17), 872–873 (2009)
23. ST. PAUL Minn: 3M RFID-based underground marking system chosen for Shanghai world exposition site. 3M News. http://findarticles.com/p/articles/mi_m0EIN/is_2008_March_6/ai_n24377165/?tag=content (2008)
24. Weinstein, R.: A technical overview and its application to the enterprise. IT Prof. **7**(3), 27–33 (2005)
25. Weis, S.A.: RFID (Radio Frequency Identification): Principles and Applications. **2**(3) (2007)
26. Wicks, A.M., Visich, J.K., Li, S.: Radio frequency identification applications in hospital environment. Hosp. Top. **84**(3), 3–9 (2006) (Heldref Publications)
27. Wu, D.L., Ng, W.W.Y., Yeung, D.S., Ding, H.L.: A brief survey on current RFID applications. In: International Conference on Machine Learning and Cybernetics (2009)

Chapter 2
RFID Security Threats and Basic Solutions

Abstract Radio Frequency IDentification (RFID) technology is challenged by numerous security and privacy threats that render the widespread of such an advantageous technology. The security threats encountered in RFID systems is different from the security threats of traditional wireless systems. This chapter is devoted to survey the existing security threats and their primitive solutions that do not consider cryptography. We classify the existing security threats into those which target the physical RFID components such as the tag, the communication channel, and the overall system threats. We discuss the physical system security solutions and the basic authentication techniques that ensure the valid identity of the communicating parties.

Like many other technologies, RFID systems confront a new set of challenges in providing security and privacy for individuals or organizations against possible threats while they are accomplishing a great productivity gains. Since the communication between the tags and the reader is performed through an unsecure wireless channel, the transmitted data is vulnerable to attacks by unauthorized readers. However, the security threats encountered in RFID systems are different from the security threats of traditional wireless systems. In this chapter, we overview the existing security threats and their primitive solutions that do not consider cryptography. We classify the existing security threats into those which target the physical RFID components, the communication channel, and the overall system threats. Then, we present the physical system security solutions and the basic authentication techniques that ensure the valid identity of the communicating parties.

2.1 Security Attacks in RFID Systems

RFID security attacks can be categorized into two main categories: privacy violations and security violations. In privacy violations, the attacker tries to harvest information from the objects by eavesdropping to the communications between the object and the reader or by tracking them. In security violations, an adversary

A. Khattab et al., *RFID Security*, Analog Circuits and Signal Processing,
DOI 10.1007/978-3-319-47545-5_2

counterfeits the behavior of a tag or a reader for making undesirable communications. Such security attacks may target the physical tag, the communication channel between the tag and the reader, or the application or the system which employs the RFID technology. Multilayer attacks also exist which affect more than one layer [10]. In what follows, we classify the existing security risks and threats according to their target into physical threats, channel threats and system threats. Of course, threats which RFID systems face today are not limited to those listed below. The characteristics of information security research is that you never know what kind of attack steps the attacker will take next. With the popularity of RFID systems, attacks targeting RFID systems will increase and become more complex.

2.1.1 Physical RFID Threats

Physical threats are those threats that use physical means to attack the RFID system to disable tags, modify their content, or to imitate them.

2.1.1.1 Disabling Tags

In these attacks, an attacker takes advantage of the wireless nature of RFID systems in order to disable tags temporarily or permanently [10]. To permanently disable a tag, the attacker may remove the tag form one item with high price and switch it with a tag of an item with low price. The other way is sending a kill command to erase the memory of the tag. Removing the antenna or giving a high energy wave to a tag will destroy the tag permanently. To disable the tag temporarily, the attacker can use a Faraday cage like an aluminum foil-lined bag in order to block electromagnetic waves from it. In other case, the attacker may prevent tags from communicating with readers by generating a signal in the same range as the reader which is called active jamming.

2.1.1.2 Tag Modification

Since most RFID tags use writable memory, an adversary can take advantage of this feature to modify or delete valuable data from the memory of the tag. This information might be critical such as the data about a patient's health which any inconsistency between the data stored on the RFID tag and the corresponding tagged object may result in serious problems. In some cases, the reader may not even notice this inconsistency during the communication and thinks that the content of the tag is unaltered.

2.1.1.3 Cloning Tags

In these attacks, the adversary clones or imitates the tags after skimming the tag's information. Each RFID tag used for identification has a unique ID number. If the ID information is exposed by the attacker, the tag can easily be copied. Now that a lot of programmable read-write tags are put into use, cloning a tag is not challenging. This new tag can then act as the ordinary tag without being detected. Such cloned tags are used in counterfeiting and spoofing system-level attack.

2.1.1.4 Reverse Engineering and Physical Exploration

To maintain the tag cost low, most RFID tags are not equipped with a tamper-resistant mechanism for an estimated long period of time. An attacker with physical access to a tag can duplicate a tag with reverse engineering, and by means of physical probing, the attacker is capable of getting confidential information stored within tag. This is different from tag cloning which does not require physical exploration of the tag. However, they also are used in counterfeiting and spoofing system-level attack.

2.1.2 RFID Channel Threats

Channel threats refer to the attacks targeting the insecure channel between a reader and a tag. Since the RFID technology uses wireless means of communication between the reader and the tag, RFID systems may face eavesdropping, snooping, counterfeiting, playback, tracking threats, and other communication security issues that lead to privacy leaks.

2.1.2.1 Eavesdropping

This threat addresses one of the main privacy concerns over the use of RFID technology. Eavesdropping happens when the channel is overheard secretly by an attacker to retrieve information from it [16]. Since RFID systems working in UHF covers more reading distance than other frequency bands, this threat is more likely to happen in it. Eavesdropping is a feasible threat and hard to be detected since it can be carried out at longer range on the communications between a tag and a valid reader while the adversary is passive and do not send out any signal (Fig. 2.1). This threat becomes serious when sensitive information is exchanged on the channel like data of a credit card without any encryption to protect them.

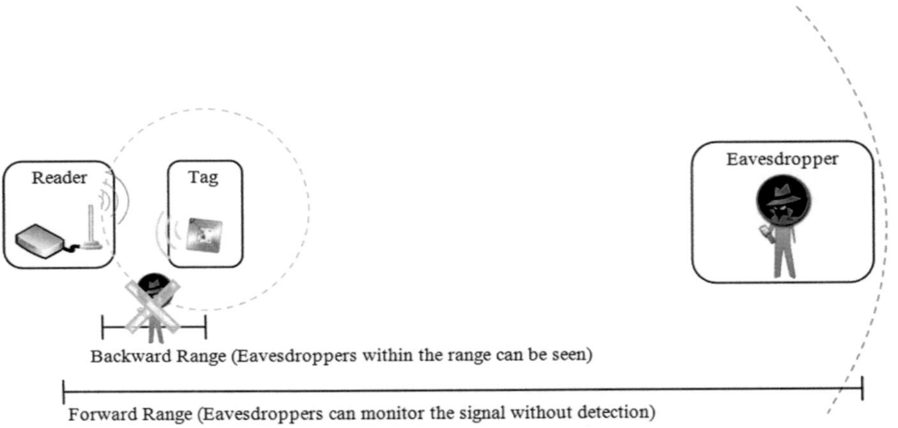

Fig. 2.1 Eavesdropping attack adapted from [2]

2.1.2.2 Snooping

This attack is defined as the illegal reading of a device's identity and data. Snooping is similar to eavesdropping with the following difference. In eavesdropping, the attacker collects the information exchanged between a legitimate tag and legitimate reader. While snooping occurs when the data stored on the RFID tag is read without the owner's knowledge or agreement by an unauthorized reader interacting the tag. This attack happens because most of the tags transmit their stored data in their memory without requesting any kind of authentication.

2.1.2.3 Skimming

In this attack, the adversary observes the information exchanged between a legitimate tag and legitimate reader. Via the extracted data, the attacker attempts to make a cloned tag which imitates the original RFID tag. To perform this attack, the attacker does not need to have any physical access to the real tag. Skimming attack is precarious when documents like drivers' licenses or passports are authenticated through RFID system. In these situations, the attackers observe the interactions between the RFID tag embedded in the document with the reader to make a fake document.

2.1.2.4 Replay Attack

One of the most serious threats which RFID systems face is the replay attack. The replay attack is when a malicious node or device replays those key information which is eavesdropped through the communication between reader and tag, in order to achieve deception. A typical application is when the illegal device playback the authentication between the reader and the tags, deceiving readers or tags to pass verification. Solutions to replay attacks include the use of stamp program, a one-time password and using the random number in authentication protocol, or updating the ID information dynamically. The researchers came up with a number of solutions to solve the problem of replay attacks such as David's Digital Library RFID protocol and distributed RFID interrogator [1].

2.1.2.5 Relay Attacks

A relay attack, also known as man-in-the-middle attack, is when an attacker places an illegal device between the reader and the tag such that it can intercept the information between the two nodes and then modify it or forwarded directly to the other end. The information transmitted through illegal devices will encounter some delay, and hence, these attack are called relay attacks.

A typical RFID relay attack system is described as follows: Suppose A is a legitimate reader, B is a legitimate label, and A' and B' are both illegal devices. A' and B' move close to the A and B, respectively, forwarding the communication information between A and B, making A believe that it communicate with B directly. The illegal device B' can be passed off as legitimate by palming off B. Meanwhile, the RFID system generally have limited communication distance, and hence, many security protocols are based on that the RFID readers and tags are in proximity are designed. However, in the relay attack, A' and B' can use other forms of communication, e.g., communication can be very far away, which destroys the premise that the reader and the tag are in proximity. An effective method to response to relay attacks is to use Distance Bounding Protocols. In 2005, Hancke et al. [6] proposed a distance limitation agreements using ultra-wide band radio, such that the readers and tags send bits of continuous authentication information to each other. By detecting the response time, the system ensures that the distance between readers and tags are closer. Later, Avoine and Reid et al. improved Hancke's agreement, achieving better results. Meanwhile, Fishkin et al. [4] found that the reader's signal to noise ratio is directly related to the distance between the reader and tag, which can be used for distance authentication.

2.1.2.6 Electromagnetic Interference

RFID channels can be the target of an adversary which aims at sabotaging the communication channel to prevent the tags from communicating with the reader. Such a communication channel threat can be either unintentional (passive interference) or intentional (active jamming).

- **Passive Interference:** Considering the fact that RFID systems operate in an inherently unstable and noisy environment, their communication is rendered susceptible to possible interference and collisions from any source of radio interference such as noisy electronic generators and power switching supplies. This interference prevents accurate and efficient communication between the tags and the readers.

- **Active Jamming:** Although passive interference is usually unintentional, an attacker can take advantage of the fact that an RFID tag listens indiscriminately to all radio signals in its range. Thus, an adversary may cause electromagnetic jamming by creating a signal in the same range as the reader in order to prevent tags from communicating with readers.

2.1.3 System Threats

System threats mainly refer to the attacks on the flaws existing in the authentication protocol and encryption algorithm. The following attacks are the main RFID system attacks

2.1.3.1 Counterfeiting and Spoofing Attacks

When the attackers get some information about the identity of RFID tags either by detecting the communication between readers and legitimate tags (skimming threats) or by physical exploration of the tags, the attacker can clone the tags. The RFID system will then be accessed using this information of identity to impersonate the legitimate labels or readers, which is called the counterfeiting or spoofing attacks. An attacker can fake labels, as well as readers. The effective means to prevent counterfeiting and spoofing attacks is to use efficient two-way authentication protocol to realize mutual authentication between tags and readers.

2.1.3.2 Tracing and Tracking

These threats violate the concept of location privacy. Illegal tracing and tracking occurs because RFID tags design requires the tag to always respond to the reader's query [16]. By sending queries and obtaining the same response from a tag at various locations it can be determined where the specific tag is currently and which locations it has visited. Since each RFID tag is affixed to a particular physical item with a unique ID number, this infers that the tag has visited those locations is which object. Encrypting the response can prevent having unauthorized access, since the adversary cannot obtain the tag contents without the secret key. However, since the tag always returns a constant response to the queries, the adversary can use this fact to perform illicit tracing and tracking.

2.1.3.3 Password Decoding

As currently most RFID systems use encryption technology to ensure the confidentiality and integrity of information delivery, attacking against the encryption algorithm is a common form of attack. Attackers can decode the encryption algorithms by conducting violent attacks, and decipher the intercepted cryptograph to get the plain-text. To respond to this attack, one need to design stronger encryption algorithms, or use longer keys to increase the difficulty of password cracking. Because of the constraint of the limited resources of RFID tags, traditional encryption or signature algorithms are difficult to be integrated into the tag. For this reason, many international scholars work on low-cost RFID encryption algorithm. For example, Yüksel proposed a low-cost 64-bit Hash function, only 1700 equivalent gates are required for the realization [18]. The Feldhofer, proposed a 128-bit Advanced Encryption Standard (AES) algorithm which requires only 3500 equivalent gates to be achieved [3], the algorithm is by far known the lowest cost AES program. The AES will be discussed in details in the next chapter.

2.1.3.4 Denial of Service (Dos) Attacks

RFID systems also may be subject to Denial of Service (DoS) attacks, which causes the system to not work properly. The attacker targets to block the reader from reading tags by using a blocker tag. Denial of service attacks are the threat to all modern communication systems. A set of mature anti-DoS solutions has developed for such threats. However, many of these solutions cannot be used in RFID systems due to the limited resources of RFID tags. For the RFID system to prevent denial of service attacks is still an area to be studied. Modern readers use anti-collision algorithms to support serving tags within their coverage areas. There are two main anti-collision algorithms; slotted ALOHA, or binary search tree. In the slotted ALOHA, the blocker tag sends an invalid packet at each time slot which will cause collision at all time slots. In binary search tree, the blocker tag will send both logic-1 and logic-0 at each bit in the serial number. Thus, the reader will be forced to search all of the possible combinations in the binary tree (i.e. if the time identifying a one serial number is 1 ms and the serial number length is 48-bit, the reader needs 1 ms $\times 2^{48} \approx 8925$ years for searching all the binary tree!!).

2.2 RFID Security Measures and Defenses

To address the various aforementioned security threats, RFID devices had to employ various security measures designed to counter the different threats. In this section, we explore these various defense techniques employed by RFIDs [12]. Our main focus in this section is on such techniques that are applicable to simple (low cost and low power) RFIDs which have limited resources. This is because more powerful

RFIDs with more resources can employ cryptography to further increase the security of the system. Cryptography principles and how it is used in RFID system will be discussed in details in Chap. 3. In contrast, simple RFID tags are unable to perform typical cryptographic operations since such simple tags has a couple of thousand gates. These gates are mainly for basic operations and only very few gates are available for use to implement security functions. The lack of computational resources is counted as a temporary state of affairs, in the hope that Moore's Law will soon render inexpensive tags more computationally powerful. However, the cost factor is still a problem since RFID are used in vast numbers. Since RFID tags replace barcodes on individual items, they will contribute substantially to the cost of those items if the tag cost is high. Hence, this section discusses security and privacy defense mechanisms that employ simple measures such as tag-killing, tag-blocking, re-encryption and many others. We classify such techniques to those which address the privacy concerns and those which address the security concerns.

2.2.1 Physical Solutions for RFID Privacy Protection

To protect the privacy of RFID tags against possible attacks and threats, physical solutions that tackle the RFID itself are helpful. In this section, we introduce such defenses and investigate their pros and cons.

2.2.1.1 Killing Tags

In this method, the RFID tags are "killed" upon purchase of the tagged product by a customer. After killing the tag, it is no longer functional and cannot be re-activated anymore. This approach is performed by sending a special command including a short password [15]. For instance, in a supermarket, the tags of purchased goods would be killed at checkout for protecting the privacy of consumers. Therefore, none of the purchased items would contain alive RFID tags.

The advantage of this solution lies in the simplicity and effectiveness of the method. However, since in this method the tag cannot be reused, its lifetime is limited and it cannot be utilized for after-sale purposes while consumers may wish to keep them alive after buying them. For example, a smart fridge which keeps the expiration dates of groceries from their tags. Based on this information, it can also give a report of what is inside it and generate a list of shopping list. Other examples of RFID tag applications include theft-protection of belongings and wireless cash cards. In these applications, the RFID tag is required to be alive when the customer buys it and it cannot be killed.

2.2.1.2 Sleeping Tags

The "sleeping" mechanism is another type of physical solutions [2]. In this approach, the reader sends a "sleep" command including a password to the tag to make it temporarily inactive. This method is similar to the killing tag method with the difference that the sleeping tag can wake up and be activated as soon as it receives the command from the reader. Meanwhile, the tag can never be re-activated in the killing tag method.

The sleeping tag approach offers an advantage to the user to switch the state of the tag between active and inactive. The problem of using this method is the existence of the possibility that the password used for controlling the tags might be overheard by an eavesdropping attack.

2.2.1.3 Faraday Cage

Faraday cage is an easy way of protecting an RFID tag that is inspired by the characteristics of electromagnetic fields and was introduced in [5]. A Faraday cage is an enclosure design made of conducting materials to exclude electromagnetic fields. Since any exterior radio signals cannot penetrate inside the cage, no reader can have access to the tag to read it as long as the RFID tag is inside such a cage.

Figure 2.2 shows how a Faraday cage shield enclosed tag from unwanted electromagnetic waves. The electromagnetic field pushes electrons of the cage toward the left. It leaves a negative charge on the left side and a positive charge on the right side of the cage. The result is that the electric field inside the cage is zero.

Faraday cages are extremely effective at providing consumer privacy against eavesdropping and tracking attacks. However, the main drawback of using this cage is its impracticality. The tag is protected from being read by unauthorized reader only when it is inside the cage. It might be practical for some items like smart cards, while using the cage is not convenient for a variety of objects like for tags injected under the skin or tags attached to a dress when it is being worn. The other problem

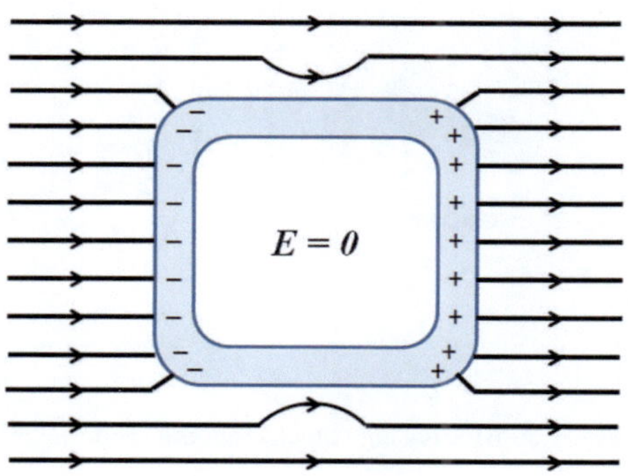

Fig. 2.2 A Faraday cage in an electric field

is preventing being read by the authorized readers unless the tag is outside the cage. Besides, using a Faraday cage for each tag imposes extra cost. These disadvantages put some limitations on using this approach which make this solution only suitable for some particular applications.

2.2.1.4 Blocker Tags

A blocker tag is a physical solution for protecting privacy in RFID systems introduced in [9]. A blocker tag is similar to an RFID tag with the difference that it can block readers from reading the identification of those tags that exist in the blocker tag's range.

The operation of blocker tags is based on creating collision for a reader when it is attempting to identify tags in its field. To identify a tag from other tags, a reader sends a query asking its serial number. Since there is a possibility that multiple of tags exist in the reader's range and respond to this query at the same time, the probability of jamming to occur is high. Therefore, readers use some algorithms like tree walking to resolve this collision. In this algorithm, each time the reader asks that only those tags which serial number starts with a special number answer. If the reader still receives more than one response, it will continue by limiting the range of serial number until just one tag answers the query. The blocker tag uses this feature and by answering all queries that reader broadcast, it fabricate a fake collision (Fig. 2.3). Thus, the reader is tricked into believing that all tags in its field are in interrogation zone. This way, a blocker tags can establish a safe zone around the tags and all RFID tags that exist in this zone can impede reading their data at the presence of a blocker tag.

One of the practical and attractive applications for blocker tags is their use in supermarkets. Before purchasing the goods, their RFID tag can be read inside the

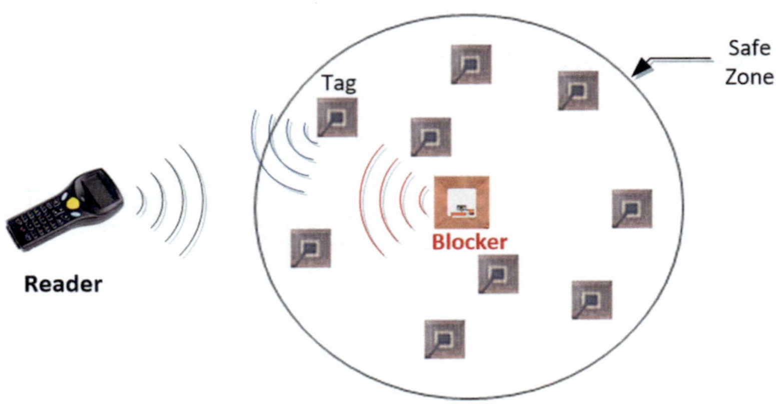

Fig. 2.3 Blocker tags blocks reading by broadcasting signals for every reader's query

supermarket without any restrictions. When they are placed in the hands of the customer, a blocker tag might be added to the shopping bag to block all further communications. This blocker tag guarantee the customer's privacy against any threats until the items are removed from the shopping bag. Then, the tags of the purchased items can operate again like before.

The major advantage of this approach is keeping the functionality of tags. Unlike killing tags wherein the lifetime of the tags are limited by the purchasing time, this method allows the tags to be more useful by expanding their lifetime. However, a major drawback of this method is its limited safety. The attacker cannot have access to tags just in a defined range and beyond this range, tags are not protected from attacks. Besides, blocker tags are not applicable everywhere. For example, in supply chains, tags are required to be available all the time and they cannot be blocked from being read by readers while the blocker tags imped all readers to have communications with tags even authorized readers.

2.2.1.5 Tag Relabeling

It is an approach in which the unique identifier of the tag is relabeled with a new unique identifier. However, the old identifier remains on the tag for further use. There are various works done based on this idea such as [17] which proposed the idea of rewriting a new random number on the RFID tags on each checkout. The authors used such a technique to present a solution for clandestine scanning of library books. Alternatively, the authors of [7] suggest two approaches for RFID tag privacy. The first tag-labeling privacy solution is based on masking the permanent ID of the tag under a private ID that is given by the users. In the other approach, the tag's permanent ID is split into two parts: a partial ID sequence that is assigned to an object, and the rest of the ID is given by user-assignable RFID tags. According to these approaches, the users have the control over the ID's uniqueness either locally or globally. Hence, the users can enable the tag's private or public ID in the different stages of the life cycle of the object.

2.2.1.6 Minimalist Cryptography

"Minimalist cryptography" in RFID tags achieves the goals of cryptography under the special resource constraints imposed by RFID tags. A "minimalist" system in which the main idea is to apply pseudonyms to help enforcing privacy in RFID tags was first proposed in [8]. In a nutshell, a tag may carry multiple, random-looking names. Each time it is queried, the tag releases a different name. In principal, only a valid verifier can tell when two different names belong to the same tag. Of course, an adversary could query a tag multiple times to harvest all names so as to defeat the scheme. This approach involves some special enhancements to help preventing such adversary. First, tags release their names only at a certain (suitably slow) prescribed rate. Second, pseudonyms can be refreshed by authorized readers. The minimalist scheme can offer some resistance to corporate espionage, like clandestine scanning of product stocks in retail environments. A new security model for EPC G2 tags which is based on minimalist cryptography was proposed in [13]. Such a model provides a solution against spoofing, replay, denial-of-service, traffic analysis and tracking.

2.2.1.7 Proxy Privacy Devices

Generally RFID readers and tags cannot have the ability to provide consumer privacy protection. One way to overcome this challenge is to rely on the reader for privacy protection. However, relying on the reader for privacy is risky due to the fact that the reader is public. Alternatively, privacy-enforcing devices can be added to RFID systems. Along with this approach, researchers have proposed several systems such the *RFID Guardian* proposed in [14]. The RFID Guardian is a platform that offers centralized RFID security and privacy management for individual people. It is integrated with four separate security policies, i.e. auditing, efficient key management, access controls and act as mediator between the RFID readers and the RFID tags as an RFID firewall.

2.2.2 Authentication

Authentication is a process through which an object proves its claimed identity to other communication party with providing some evidence such as what it knows, what it has, or what it is. This process is applicable through only software solutions and it is not possible by physical solutions. In RFID systems, authentication is required in two phases. First, before beginning any communication, both the tag and the reader should verify their identity to make sure that they are contacting with the wished partner. The second phase is when data is exchanged between the two parties to ensure that the exchanged data is intact.

When a tag passes through the electromagnetic field of a reader, it becomes activated and can detect the reader's signal. To reply to the reader, the tag needs to know if the reader is the legitimate one or not. Otherwise, an unauthorized reader can obtain information from tags which are currently in its field by eavesdropping and keep a tracking of their current locations. Also, an unauthorized reader can have access to the tag's memory to read or even manipulate its data. Therefore, to prevent these threats, a process is required to authenticate the reader to the tag. On the other hand, the reader is required to find out if the tag contacting with is reliable or not. This way, the reader can make sure that it is not communicating with a counterfeit tag. This process is called authenticating tag to the reader. Mutual authentication permits the two parties to authenticate each other's identity. This happens when both tag to reader authentication and reader to tag authentication

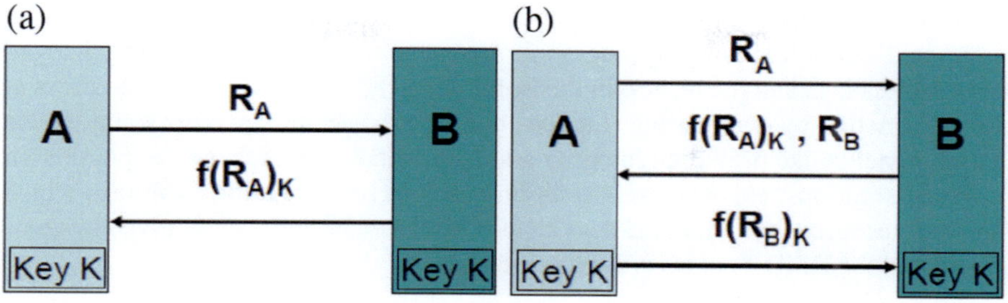

Fig. 2.4 Challenge-response technique in symmetric authentication. (**a**) Unilateral authentication. (**b**) Mutual authentication [11]

are performed. Conducting mutual authentication between RFID tags and readers should be performed before exchanging any key and data. This way, all of the former mentioned security problems in the last sections can be solved.

Implementing unilateral and mutual authentication at the beginning of the communication has been the focus of many researches. The authors of [11] presented three authentication methods. The first method, password authentication, provides a weak level of security. Customized and zero-knowledge authentication is another technique based on mathematical problems, the implementation of which imposes high cost. Challenge-response is a high secure scheme which is being of interest recently. This scheme is categorized into two groups: symmetric and asymmetric. Asymmetric techniques are time consuming and their implementation cost is high. On the contrary, symmetric methods need key exchange and management since they use one shared secret key (Fig. 2.4).

During communication, providing authentication is required since there is a possibility that attackers send the message on behalf of each party or manipulate the message such that they replace their desired message with the real one. This service can be implemented by keyed hash function or Message Authentication Codes (MAC). Using MACs bring the benefit that the integrity of the message can be guaranteed. Authentication is essential when the possibility of existing attackers are high like battle fields or the condition of environment is harsh and may affect the accuracy of the messages. Also, performing this service is vital in applications in which the value of data is important such as health care applications.

2.3 Concluding Remarks

Considering the limitations and drawbacks of the physical solutions discussed in this chapter for providing security and privacy in RFID applications, these solutions are suitable for particular applications and cannot be applicable for all applications. Other solutions are required that does not suffer any limitation on the life-span of tags such as in killing method or block authorized readers like faraday cage.

Such solutions also should not be restricted to a special zone like blocker tags. The suggested solution is using cryptographic algorithm to encrypt messages exchanged between the tags and the reader. In this solution, an adversary cannot have access to the information by overhearing if it does not have the secret key. This solution also brings benefits like providing integrity and authentication which are not possible in physical solutions. However, this solution needs to be compatible with tags which are very resource limited. In the next chapter, a survey of lightweight cryptosystems developed for RFID systems will be presented.

References

1. Chauhan, M., Sharma, E.: A survey on RFID technology. Int. J. Res. **1**(10), 1316–1322 (2014)
2. Chen, Y., Tsai, M.: The Study on Secure RFID Authentication and Access Control. InTech (2011)
3. Feldhofer, M., Dominikus, S., Wolkerstorfer, J.: Strong authentication for RFID systems using the AES algorithm. In: Cryptographic Hardware and Embedded Systems-CHES, vol. 3, pp. 357–370. Springer, Berlin (2004)
4. Fishkin, K.P., Roy, S., Jiang, B.: Some methods for privacy in RFID communication. In: Security in Ad-hoc and Sensor Networks, pp. 42–53. Springer, Berlin (2005)
5. Garfinkel, S., Rosenberg, B.: RFID: Applications, Security, and Privacy. Addison-Wesley, Reading, MA (2006)
6. Hancke, G.P., Kuhn, M.G.: An RFID distance bounding protocol. In: Proceedings of IEEE 1st International Conference on Security and Privacy for Emerging Areas in Communications Networks [SecureComm 2005] (2005)
7. Inoue, S., Yasuura, H.: RFID privacy using user-controllable uniqueness. In: Proceedings of RFID Privacy Workshop (2003)
8. Juels, A.: Minimalist cryptography for low-cost RFID tags. In: Proceedings of 4th International Conference on Security Communication Networks. Lecture Notes in Computer Science, vol. 3352, pp. 149–164. Springer, Berlin (2004)
9. Juels, A., Rivest, R.L., Szydlo, M.: The blocker tag: Selective blocking of RFID tags for consumer privacy. In: Proceedings of the 10th ACM Conference on Computer and Communications Security, CCS '03 (2003)
10. Mitrokotsa, A., Rieback, M., Tanenbaum, A.: Classifying RFID attacks and defenses. Inf. Syst. Front. **12**(5), 491–505 (2010)
11. Menezes, A.J., van Oorschot, P.C., Vanstone, S.A.: Handbook of Applied Cryptography. CRC Press, Boca Raton (1997). Availableonlineathttp://www.cacr.math.uwaterloo.ca/hac
12. Pateriya, R.K., Sharma, S.: The evolution of RFID security and privacy: a research survey. In: IEEE International Conference on Communication Systems and Network Technologies [CSNT] (2011)
13. Qingling, C., Yiju, Z., Yonghua, W.: A minimalist mutual authentication protocol for RFID system & BAN logic analysis. In: Proceedings of ISECS International Colloquium on Computing, Communication, Control, and Management (2008)
14. Rieback, M., Crispo, B., Tanenbaum, A.: RFID guardian: a battery-powered mobile device for RFID privacy management. In: Proceedings of Australasian Conference on Information Security and Privacy. Lecture Notes in Computer Science, vol. 3574, pp. 184–194. Springer, New York (2005)
15. Sarma, S., Weis, S., Engels, D.: RFID systems and security and privacy implications. In: Cryptographic Hardware and Embedded Systems - CHES 2002. Lecture Notes in Computer Science, vol. 2523, pp. 454–469. Springer, Berlin (2003)

16. Weis, S., Sarma, S., Rivest, R., Engels, D.: Security and privacy aspects of low-cost radio frequency identification systems. In: Security in Pervasive Computing. Lecture Notes in Computer Science, vol. 2802, pp. 201–212. Springer, Berlin (2004)
17. Wu, D.L., Ng, W.W.Y., Yeung, D.S., Ding, H.L.: A brief survey on current RFID applications. In: International Conference on Machine Learning and Cybernetics (2009)
18. Yüksel, K.: Universal hashing for ultra-low-power cryptographic hardware applications. Ph.D. thesis, Worcester Polytechnic Institute (2004)

Chapter 3
Cryptography in RFID Systems

Abstract To provide security and privacy in RFID systems, physical solutions are not suitable because of their limitations and disadvantages. Instead, cryptography is an inevitable way to make the RFID technology secure. From a theoretical point of view, standard cryptosystems might be an accurate approach. However, they demand resources far more than those available to many tags in terms of circuit size, power consumption and area. Since low-cost RFID tags are very constrained devices with severe limitations in their budget, lightweight cryptographic techniques are the most appropriate solution for such RFID tags. In this chapter, the characteristic of a lightweight cryptosystem are defined. Then, a set of the well-known and most recent lightweight cryptography implementations is presented. This survey covers the recent hardware implementations of symmetric as well as asymmetric ciphers.

One of the main challenges that face the Radio Frequency Identification (RFID) technology is its vulnerability to security attacks by unauthorized reader(s) which can interrogate or modify the information stored in the tags. Several encryption solutions have been developed for wireless communication systems to address such security challenges. On the one hand, there exist several asymmetric or public key encryption algorithms that use two keys to secure data in networked systems. However, such solutions are not applicable to RFID systems—despite their high security performance—due to the limited processing and power capabilities of the tags. Even existing highly-optimized hardware implementation of such algorithms are way beyond what a typical RFID system can afford. Recall that RFID encryption algorithms must be light enough in terms of area and power to satisfy the resource limitations of RFID systems. Likewise, using hash functions is not suitable for constrained environments since they require significant amounts of resources in their designs, and hence, they are not hardware friendly.

On the other hand, several symmetric or private key encryption algorithms have been developed which are less resource hungry compared to public key encryption algorithms. As will be discussed in detail in this chapter, existing private key security algorithms promise reasonable security and meet the low resource requirements of RFID systems. However, the main drawback of existing algorithms is that they do not provide integrity and authentication services along with confidentiality. Integrity is very important service for harsh environments wherein the possibility

A. Khattab et al., *RFID Security*, Analog Circuits and Signal Processing,
DOI 10.1007/978-3-319-47545-5_3

of corrupting the received data is high. In hostile environments, authentication is a must in order to ensure access of the RFID information only by the authorized readers. In order to provide integrity and authentication alongside confidentiality, existing symmetric algorithms are required to be integrated with other algorithms (such as message authentication code (MAC) algorithms) in order to provide the targeted authentication and integrity services. Such integration results in a high cost in terms of power and area which cause such symmetric key algorithm to lose their main advantage.

In this chapter, we first define the characteristic of a lightweight cryptosystems. Then, well-known and most recent lightweight cryptography implementations are presented. This survey covers the hardware implementations of symmetric as well as asymmetric ciphers.

3.1 Wireless Security Preliminaries

The blast of the Internet along with its exponentially growing number of users and applications, such as RFID systems, resulted in creating a new world referred to as cyberworld. This virtual world brings dependency on electronic systems, sharing data globally, and rising device connectivity and online communications which make new meanings for business, education, sociality, and entertainment in our society.

Although the cyberworld offers the ordinary users extraordinary capabilities by building a digital extension of the real world and map it to a virtual environment, it opens potential opportunities for abuse and crimes against users' privacy as well. Cybercrimes include committing activities like intrusion attempts, access to unauthorized data, denial of service attacks, identity thefts, digital fraud, and data tampering [33, 64]. To protect data and systems from cyber-criminals the following four requirements are essential:

Confidentiality: Only the sender and the intended recipient of a communication can see the content of that communication. This concept is accomplished through encryption.

Data Integrity: It guaranties that the data received at the receiving party is original and was received exactly as it was sent by the sending party. If the content of a communication is compromised, it must be detectable by either of the communicating parties. Data integrity can be threatened either by environmental hazards—such as heat, dust, and electrical surges—or by attackers.

Authenticity: The sender and the receiver should be able to verify each other's identity. Any impostor needs to be either detected or identified.

Non-repudiation: It means preventing an entity from denying previous actions. In other words, the sender of the message cannot deny having sent the message.

Among these four services, confidentiality is the primary service and all security algorithms are required to provide it, while other services are arbitrary.

3.2 Cryptography Overview

Cryptography is a fundamental method for ensuring the above mentioned requirements [30]. The word "cryptography" is derived from the Greek roots, "kryptos" and "graphein" meaning secret writing. Cryptography is studying different techniques concerned with keeping the communication between two parties private in the presence of third parties. An encryption scheme is composed of five components: a plaintext, an encryption algorithm, a secret key, a ciphertext, and a decryption algorithm. In these techniques, a message, called plaintext will be converted at the sending party using a secret key and an algorithm or a mathematical procedure such that the result, called ciphertext, appears non-sense for all parties. The used algorithm for encryption and decryption is available for all parties while the secret key is shared only between the sender and the receiver. Recovering the plaintext from the ciphertext, called decryption, is possible for those parties who have access to the right key. Good cryptographic algorithms should always be well designed such that they are not easy to break. This can be achieved by fulfilling the above four requirements: confidentiality, data integrity, authenticity, and non-repudiation.

Cryptography algorithms could be classified into three categories: private key algorithms, public key algorithms, and hash functions.

3.2.1 Symmetric Private Key Encryption

Private key algorithms, also called symmetric encryption algorithms, are the oldest cryptographic techniques used to have secure communications. In these algorithms, the same key is used for both encryption and decryption as shown in Fig. 3.1. An encrypted message is considered to be confidential if only those parties that have the shared secret key can recover the plaintext.

Private key algorithms are acceptable solutions for many applications especially when there are resource limitations. The following are some of the well-known private key encryption algorithms:

- Data Encryption Standard (DES) that is designed specifically to yield fast hardware implementations and slow software implementations [13].
- Triple DES (3DES) which is a variant of DES that makes three encryption/decryption passes over a data block [14].
- Advanced Encryption Standard (AES) [11] that is the official successor to DES.
- Blowfish that is optimized for 32-bit processors with large data caches [50].
- Secure and Fast Encryption Routine (SAFER) that is designed for efficient software implementation [5].
- Welch-Gong (WG)-based stream cipher (WG-8) which software implementation is optimized for microcontrollers [23].

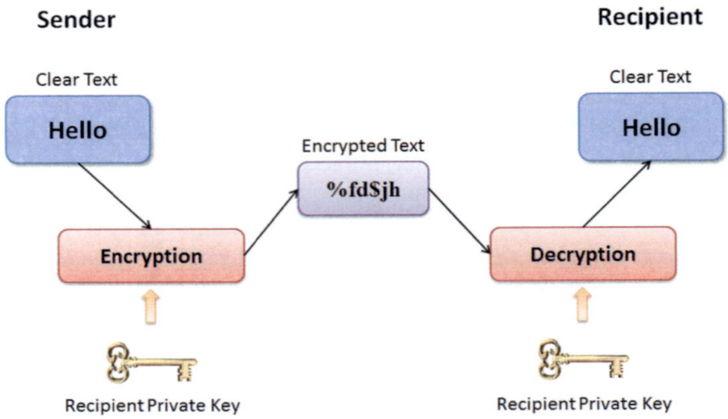

Fig. 3.1 Symmetric private key encryption

Although there are efficient software and hardware implementations for private key algorithms, these algorithms have some drawbacks. Private key algorithms are not capable of providing authentication and integrity on their own and they need other algorithms to be integrated with them for supporting these services. Besides, these algorithms are not able to provide non-repudiation service. It means that a third party cannot prove who really has sent a message. Moreover, key management and distribution among users could be an overwhelming task as both sides of the communication should have access to the secret key. These are the main drawbacks of private key encryption algorithms compared to public-key encryption algorithms discussed in the following sub-section.

3.2.2 Asymmetric Public Key Encryption

Public key algorithms depend on the existence of the so-called one-way functions which are mathematical functions that are easy to compute whereas their inverse function is relatively difficult to compute such as exponentiation versus logarithms. For example, computing $3^6 = 729$ is easy whereas finding x and y such that $\log_x 729 = y$ is very difficult.

Until the middle of the 1970s, the encryption systems were based on private key algorithms. The idea of public key encryption was first introduced by Diffie and Hellman in 1976 [15]. In Public key encryption algorithms, there are two different keys: one for encryption and the other one for decryption as shown in Fig. 3.2. Since these two keys are different, these algorithms are also called asymmetric encryption algorithms. One of these keys, called the public key, is published in the network and is used to encrypt the plaintext at the sender part. The other one, called the secret key, is kept private and is used to decrypt the ciphertext at the receiver party.

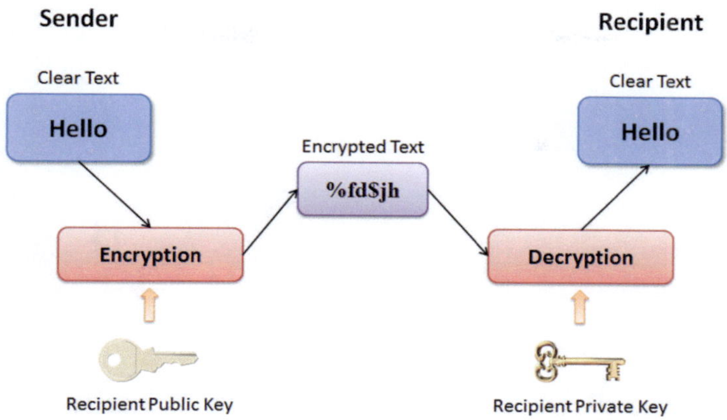

Fig. 3.2 Asymmetric public key encryption

An important property for these algorithms is the difficulty of finding the secret key while everyone knows the public key. Therefore, everyone can encrypt a message with a particular party's public key but just that particular party can decrypt that message. This way, no one else can recover the message. This technique has resolved the discussed problems in symmetric algorithms. Since the public key is published and it is not required for all parties to keep the encryption key in private, no key is required to be exchanged among the involved parties. Besides, these two keys are interchangeable, meaning that the sender can encrypt the plaintext with its own private key and the receiver can decrypt it with the sender's public key. This property is used to provide authentication and non-repudiation services.

The most popular and most widely used public-key cryptosystems are RSA introduced by Rivest, Shamir and Adleman in 1977 [61] and Elliptic Curve Cryptography (ECC) proposed independently by Koblitz and Miller in 1985 [39, 48]. The security strength of RSA relies on the fact that the factorization of large numbers is difficult. ECC algorithm is based on discrete logarithm problems on elliptic curve groups. Other public-key cryptosystems have been also proposed such as ElGamal [19] Rabin [55] and NTRU [32].

Public key algorithms are extremely secure compared to private key algorithms. However, their implementations are much more complex as well. As a result, their computation speed is relatively poor. Furthermore, although speed up through hardware implementation is possible, public key hardware systems use more die space and usually require more power than private key systems. In summary, since public key algorithms often rely on complicated mathematical computations, they generally are much more resource hungry compared to private key algorithms in resource restricted applications.

Fig. 3.3 Hash function

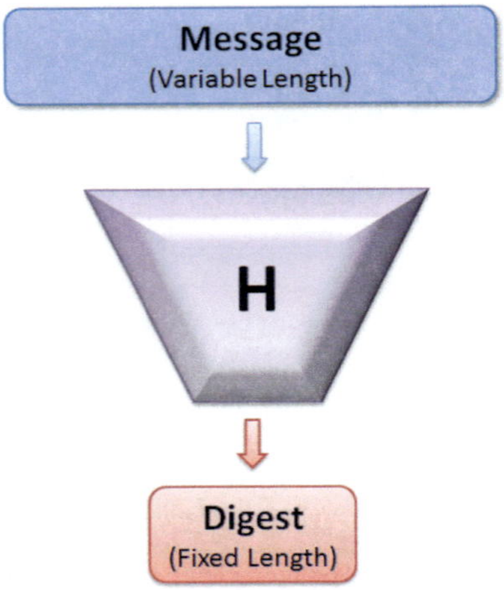

3.2.3 *Hash Function*

A hash function takes a block of data, called the message, and returns a fixed-size output, called the hash value or digest as shown in Fig. 3.3. A small change in the input data will change the digest completely.

Unlike private key and public key algorithms, hash functions are one-way encryption. This means that it is easy to compute the output for every input but it is impossible to find either the input for a given output or even the length of the input. This property of hash functions can be useful in security applications like digital signature, message authentication codes (MACs) and authentications. The main application of hash functions in cryptography is message integrity. The hash value provides a digital fingerprint of a message's contents, which ensures that the message has not been altered by an intruder, virus, or by other means. Hash algorithms are effective if there is extremely low probability that two different plaintext messages yield the same hash value. There are some other applications for hash functions such as password verification, regenerating keys from a single key, and file or data identifiers.

The ideal cryptographic hash function is required to have four significant properties:

- It is easy to compute the hash value for any given message.
- It is impractical to generate a message that has a given hash.
- It is impractical to modify a message without changing the hash.
- It is impractical to find two different messages with the same hash.

The first and second properties support the definition of the hash function. The third property, called week collision, and the forth property, called strong collision, will protect the security of the hash function against attackers who try to substitute their own message with the original message.

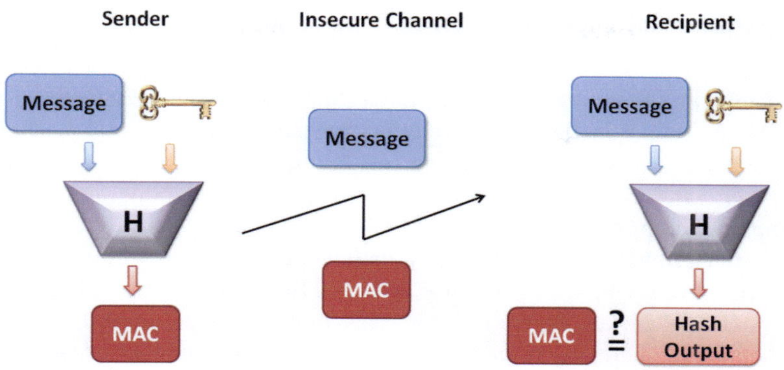

Fig. 3.4 Using hash function for verifying the received message

Hash functions can also be combined with other standard cryptographic methods to verify the source of the data. When hashing algorithms are combined with encryption, they produce special message digests that identify the source of the data. These special digests are called Message Authentication Codes (MAC). This process is illustrated in Fig. 3.4. This process, also called keyed hash function, accepts a message plus a secret key. The output protects both message integrity and message authentication.

There are several well-known hash functions in use today:

- Hashed Message Authentication Code (HMAC): Combines authentication via a shared secret with hashing [4].
- Message Digest 2 (MD2): It is byte-oriented and produces a 128-bit digest which is designed for smart cards [36].
- Message Digest 4 (MD4): It is very similar to MD2 and is designed for fast processing in software [59].
- Message Digest 5 (MD5): It is similar to MD4 but slower because the data is manipulated more [60].
- Secure Hash Algorithm (SHA): Produces a 160-bit digest. It is modeled after MD4 and proposed by NIST for the Secure Hash Standard (SHS) [17].

Traditionally, private key encryption, public key encryption, and hash functions have been used together in real-life implementations. The public key systems are used to initiate a communication and then the majority of the communications are done using private key encryption. Hash functions are also used for data integrity. This strategy, called hybrid cryptosystem, combines the feasibility of public-key cryptosystems with the efficiency of symmetric-key cryptosystems. The public key algorithm is used to authenticate the users and exchange the secret session key at the beginning. Since this is only a small portion of the entire communication, speed penalty of using public key encryption will be compensated by the enhanced speed offered by private key encryption in the later part of the communication.

3.3 Lightweight Cryptography

Lightweight cryptography is an innovative approach which aims at providing solutions to meet the challenge of developing fast and efficient security mechanisms for harsh resource constrained environments. These solutions include new designs in cryptographic primitives and protocols in addition to adapting and modifying contemporary cryptosystems [54].

To design a lightweight cryptography, there are three aspects which are required to be optimized: security, performance and cost. Security is measured through the number of bits of the key. By increasing the size of the key, the provided security will be higher. Performance is considered in terms of the total number of clock cycles to complete an operation which is proportional to the throughput and energy. The cost, e.g., expressed in terms of power or area, depends on the used architecture. Among these three aspects, there is a trade-off which makes optimizing all of them together in one design very difficult (Fig. 3.5). For example, security is in trade-off with performance and cost. Having high security requires increasing either the number of rounds or the cost. Performance and cost are two other vertexes of this triangle. Serialized architecture yields lower power and area while it results in lower performance.

To have a more precise definition of lightweight cryptography, it is required to define the boundaries of cost and performance. The power consumption of a security implementation has to be reduced to 10 s of microwatts, and for EEPROM read operation this limitation should not exceed it unless the tag read range requirements cannot be preserved [44]. A complete RFID tag, including the analog part, might have between 1000–10,000 gate equivalent (GE). For the security part of the tag,

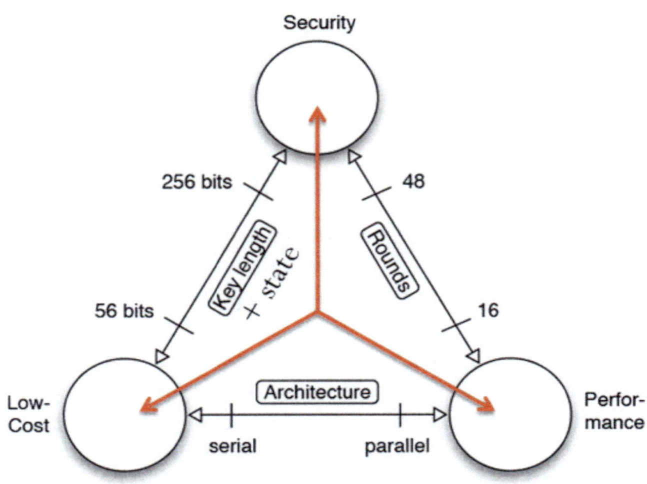

Fig. 3.5 Design trade-offs for lightweight cryptography

this margin may be kept between 200–2000 GE [34]. Performance is mainly limited by the user requirements and the air interface protocols. However, it is recommended to be 10 to 100 s clock cycles.

In the following sections, the literature of lightweight cryptography is studied. In this survey, some new lightweight design and some modified contemporary cryptosystem will be investigated separately. At the end, a comparison of these designs will be presented.

3.4 Asymmetric Key Encryption Lightweight Cryptosystems

As discussed earlier in this chapter, asymmetric key encryption algorithms, also called public key algorithms, are very strong in terms of security. They provide confidentiality, integrity, reliability, availability and non-repudiation altogether. In this type of cryptography, two different keys are used: a public key which is published on the network and a private key which is kept secret to the user as shown in Fig. 3.2. To encrypt a plaintext, a public key is enough, but to decrypt the ciphertext, the corresponding private key is required. Thus every party can encrypt a message while only the party that has the private key can recover the message. Public key constructions are typically based on some mathematical problem, such as factoring, which is assumed to be a hard problem in a computational sense. For example, in factoring, the private key can consist of two large prime numbers and the corresponding public key is their product. Obtaining the private key from the public key is possible in theory, but in practice, a huge amount of resources (e.g., time) is required to compute it.

One of the advantages of asymmetric key cryptosystems is distributing keys among parties. Since it is not required for all parties to keep the encryption key in private, no key is required to be exchanged among involved parties.

Public key algorithms are extremely secure compared to private key algorithms. However, their implementations are much more complex as well. As a result, their computation speed is relatively poor. Although speed up through hardware implementation is possible through parallelism, public key hardware systems use more die space and usually require more power than private key systems. Furthermore, since public key algorithms often rely on complicated mathematical computations, they are generally much more resource hungry compared to private key algorithms. Nevertheless, some researches have been done towards adapting public key algorithms with resource restricted applications. Next, one of the most well-known public key algorithms, ECC, is studied.

3.4.1 Elliptical Curve Cryptography (ECC)

Elliptical curve cryptography (ECC) is a public key encryption technique based on elliptic curve theory over finite fields. ECC-based systems offer similar security for smaller key sizes compared to RSA-based systems [18]. Since the computational and area complexities of the hardware implementations for cryptographic algorithms are proportional to their key sizes, ECC-based systems are smaller, faster, and consume less power compared to RSA-based systems.

In ECC cryptosystem, all parties agree on all parameters defining the elliptic curve and a base point on this curve. Each party selects a number as a private key and compute multiplication of the base point with its private key. The result of the multiplication will be another point on the curve which is published as a public key. Finding the original point from the result is very difficult even with knowing the base point. This property guarantees the security of the ECC algorithm. To encrypt a message in ECC, the sender will first compute a shared secret key by multiplying the receiver's public key with its own private key. Then the message is added to this shared key and sent out.

A lot of research has been carried on hardware-efficient ECC implementations. In [46], the authors have tried to adapt the ECC algorithm with RFID systems by reducing the number of registers, operations, the operating frequency and also using restructured formulas as much as possible in order to meet the resource limitations of RFID systems. However, their proposed hardware is still far from the boundaries of RFID systems.

Making public key algorithm lighter is another solution. Reducing the flexibility of ECC algorithm by limiting the number of parameters such as using only one special elliptic curve [43], selecting specific field sizes [41] or choosing specific prime numbers [3, 26, 37, 53] are other ways to make ECC lighter. Although applying dedicated hardware with these limitations leads to meet the power limitation, but any change in security parameters imposes the replacement of all tags with new ones. The results of the hardware implementation of these designs in Table 3.1 indicates that they are still away from the definition of lightweight cryptosystem in terms of area, performance and power despite of all improvements performed in them.

Table 3.1 Hardware implementation results for ECC

Design	Bits	Area [gates]	Tech. [μm]	Op. freq. [kHz]	Perf. [ms]	Power [μW]
Luo et al. [46]	226	16,900	0.18	1280	N/A	6.6
Kumar and Paar [41]	131	11,969	0.35	13,560	18	N/A
Gaubatz et al. [26]	100	18,720	0.13	500	410.45	<400
Batina et al. [3]	134	6103	0.13	200	210	<13
Lee et al. [43]	163	12,506	0.13	1130	244.08	36.63
Öztürk and Sunar [53]	167	30,333	0.13	20,000	31.9	990
Kaps [37]	61	18,720	0.13	500	817.7	394.4

3.5 Symmetric Key Encryption Lightweight Cryptosystems

Symmetric key encryption is the oldest and best-known technique to provide security in communications. In this technique, the sender and the receiver both share a secret key, which they have already agreed on. The shared key is used for both encryption and decryption as shown in Fig. 3.1. This setting, referred also to as private key cryptography, is considered to be confidential if only eligible parties which have access to the shared secret key can recover the plaintext from the ciphertext.

There are several drawbacks which make symmetric key algorithms less appealing to some applications. One of the obvious problems is distributing the private keys among the authorized parties. Moreover, keeping one secret key for each party makes managing keys more difficult by increasing the number of parties. Symmetric encryption algorithms cannot provide integrity and authentication on their own. To provide these services, they need other algorithms to be integrated with them. Not supporting non-repudiation service is another problem of these cryptosystems. Despite all of these drawbacks, there are efficient software and hardware implementations for private key algorithms which make them suitable for restricted resource applications. Furthermore, recent research efforts have been directed towards private key schemes since public key algorithms have still significant challenges for RFID systems' implementation.

There are traditionally two classes of symmetric encryption algorithms: block ciphers and stream ciphers. Recently, a new class, called hybrid cipher, which is a combination of these two ciphers has been introduced.

3.5.1 Block Ciphers

A Block cipher is an encryption function that works on fixed size blocks, typically 32–256 bits. For example, AES performs on blocks of 128 bits, while other block ciphers use smaller block sizes such as 64 in PRESENT [7]. Therefore, the size of the ciphertext is fixed independent of the size of the message. In general, a block of N-bit plaintext is replaced with a block of N-bit ciphertext in block ciphers. Block ciphers like DES, 3DES, AES break message into blocks. Then, each of these blocks is encrypted while the key is the same for every block (Fig. 3.6). These ciphers repeat one or more simple operations like substitution and permutation several times. The encryption process is different from the decryption process in block ciphers.

To provide confidentiality of a communication, ciphers are required to obscure the statistical properties of the original message completely by providing confusion and diffusion between the message and the key.

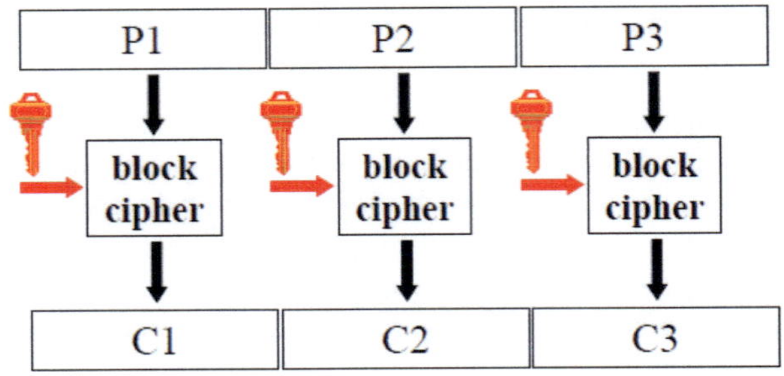

Fig. 3.6 Block cipher operations on fixed size blocks

Confusion is a way to make the relationship between the plaintext and the ciphertext as complex as possible. It can be achieved by using a complex substitution algorithm. Thus, even if an attacker can handle the statistics of the ciphertext, it is very difficult to assume the key. Caesar ciphers have poor confusion while Polyalphabetic substitutions have good confusion.

Diffusion is a way to spread the effect of changing the individual plaintext over the value of ciphertext digits as much as possible, like permutation or transposition ciphers. By globalizing the local effects, tracking the effects of each plaintext digit on the ciphertext digits will be more complicated for an attacker.

One type of modern block ciphers is substitution-permutation network or SP which is based on the two primitive cryptographic operations: substitution box and permutation box.

Substitution Box or S-Box is a basic component in block ciphers which substitutes n-bit data in the input with m-bit data in the output. Usually m and n are equal. S-Boxes are fixed look-up tables (LUT) used to provide high non-linearity and high Boolean function complexity relationship between the plaintext and the ciphertext to satisfy the confusion property in block ciphers. They use big area which makes them expensive in hardware implementation. For example, 8×8 S-Box as found in AES [25] needs 300 GE, 6×4 S-Box in DES [65] requires 120 GE and 4×4 S-Box as used in PRESENT [7] is implemented by 28 GE. In contrary, S-Boxes are suitable to be implemented by software because they can be replaced by small sized memories in software implementation. For example, for software implementation of 8×8, 6×4 and 4×4 S-Boxes, 256, 64 and 16 Byte ROM memories are required, respectively. Thus, the selected S-Box is required to be small in hardware implementation to save more cost in area. Since S-Boxes can be implemented in a single LUT, the hardware implementation of S-Boxes in FPGA is easily applicable with saving cost in area.

Permutation Box or P-Box is another helpful tool for encryption in block ciphers. It is a basic component in block ciphers which performs reordering on n-bit input to n-bit output to satisfy the diffusion property of block ciphers. It is a reversible function. Therefore, it can be used to retrain the message by the same hardware. Permutation is very suitable for hardware implementation since no gate is required and it is composed of just wiring. However, it brings complexity in routing in low level design fabrication. Since there is no gate, no transition will occur. Thus, no extra delay and power cost will be imposed. In contrary, this method is difficult in software implementation since it needs cumbersome bit operations. For example, permutation of 64 bits needs 64 cycles and 64 byte ROM memory.

Several mature block ciphers based on S-Box and P-Box are available for limited resource environments like AES and PRESENT which will be discussed in the following sub-sections.

3.5.1.1 Advanced Encryption Standard (AES)

The Advanced Encryption Standard (AES) is a symmetric key block cipher published by the National Institute of Standards and Technology (NIST) in December 2001. It is the successor of DES and an example of SP network which operates on a fixed 128-bit block of data with supporting 128, 192, or 256-bit key sizes [11]. It is organized in a 4×4 column-major order matrix of bytes, called STATE. The number of rounds in AES depends on the size of the key, e.g., 10 rounds for AES-128. To provide confidentiality, AES uses four types of transformations at each round:

- SubBytes—Each byte in the STATE matrix is replaced with a SubByte using an 8-bit Sbox (Fig. 3.7a).
- ShiftRows—Each row of the state is shifted cyclically a certain number of steps (Fig. 3.7b).
- MixColumns—Four bytes of each column of the state are combined using a linear function (Fig. 3.7c).
- AddRoundKey—Each byte of the STATE is combined with the round key using bitwise addition (Fig. 3.7d).

Among block ciphers, AES is well known block cipher for encryption. Many low-cost implementations of the smallest variant, AES-128 have been published which bring down the size of cipher to only 3100 gate equivalents [29]. An ultra-low power and low energy AES design is presented in [38]. However, the best known lightweight AES design requires 3100 gate equivalent (GE) for implementation [29], which is still significantly higher than the assumed 2000 GE. Hence, it is not a good candidate for extremely constrained device such as RFID systems. This might be due to the fact that the AES has good software implementation properties but it is not designed with hardware-friendly properties. Table 3.2 summarizes the characteristics of the different AES implementations. It is worth noting that the gate count for a hardware implementation of AES is not very likely to further decrease. Therefore, AES is often not considered as an option for developing such technology. Instead, it is considered as a benchmark for the comparison of different encryption algorithms.

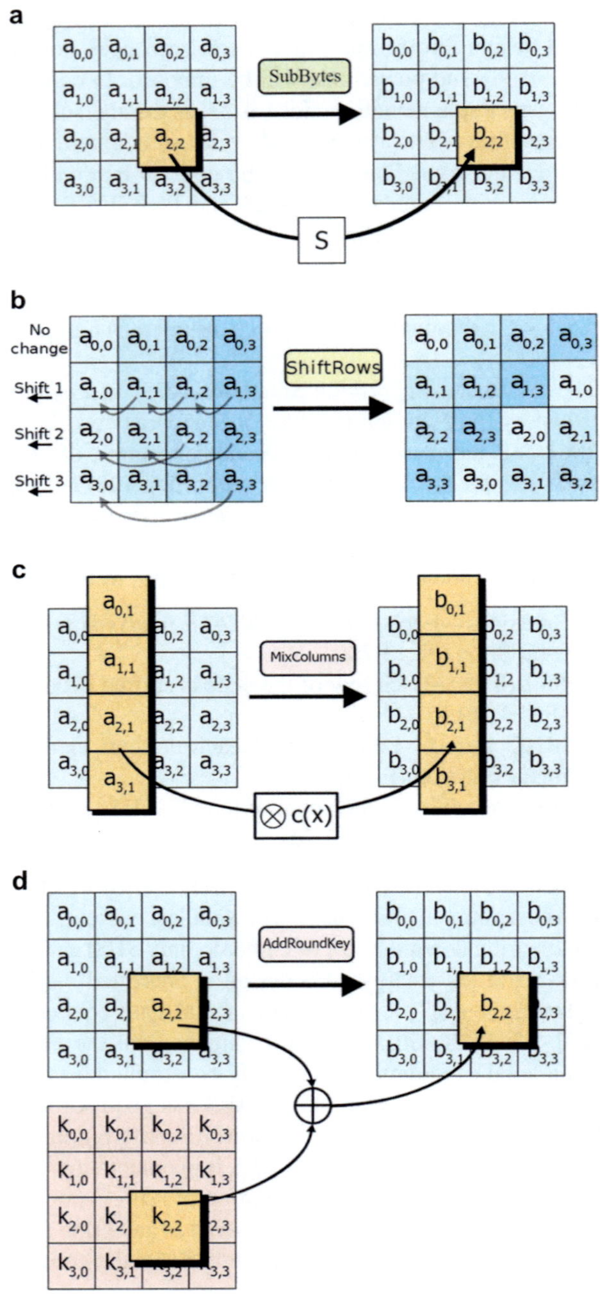

Fig. 3.7 AES four steps adapted from http://en.wikipedia.org/wiki/Advanced_Encryption_Standard. (**a**) The SubBytes step. (**b**) The ShiftRows step. (**c**) The MixColumns step. (**d**) The AddRoundKey step

Table 3.2 AES implementation characteristics

Platform	Area [GE]	Tech. [μm]	Max freq. [MHz]	Cycles per block	Throughput [Mbps]	Power [μW/MHz]	Mode
Feldhofer et al. [25]	3400	0.35	80	1032	9.9	45	En/De
Hamalainen et al. [29]	3200	0.13	130	160	104	30	En
Kaps and Sunar [38]	4070	0.13	N/A	534	N/A	47.66	En/De

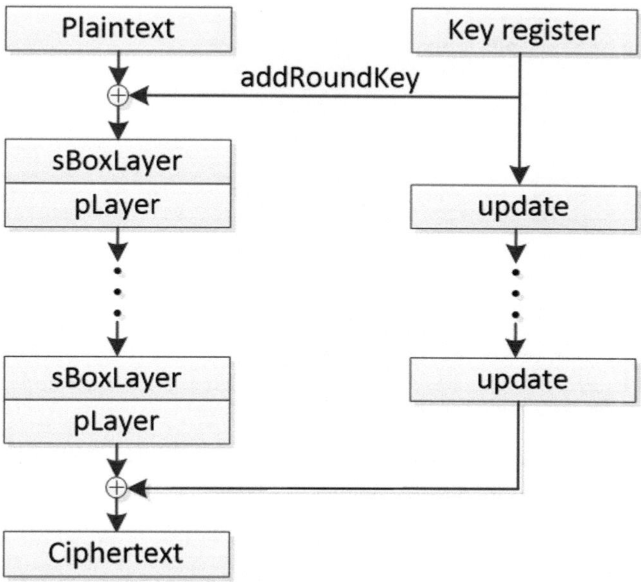

Fig. 3.8 A top-level algorithmic description of the PRESENT algorithm adapted from [7]

However, AES is known to be vulnerable to the following attacks: known-key distinguishing attack [27], chosen-key-relations-in-the-middle attacks [58], key-recovery attacks based on bicliques [6].

3.5.1.2 PRESENT

PRESENT is a block cipher based on a SP network, inspired by the techniques used in DES and AES and designed for resource-constrained systems. It operates on a 64-bit block of data, supporting 80-bit and 128-bit keys [7].

The encryption process is completed in 31 rounds. Each round is composed of three layers: addRoundKey layer, sBoxLayer and pLayer (Fig. 3.8). These three layers are followed one by one at each round. At the beginning of the encryption process, the Key register and STATE are initialized with the encryption key and plaintext, respectively. At each round, the key register is updated by rotating 61 bits

Algorithm 1 PRESENT algorithm

generateRoundKeys()
for $i = 1$ to 31 **do**
    addRoundKey(STATE,K_i)
    sBoxLayer(STATE)
    pLayer(STATE)
end for
addRoundKey(STATE, K_{32})

Table 3.3 The PRESENT S-Box [7]

X	0	1	2	3	4	5	6	7	8	9	A	B	C	D	E	F
$S(x)$	C	5	6	B	9	0	A	D	3	E	F	8	4	7	1	2

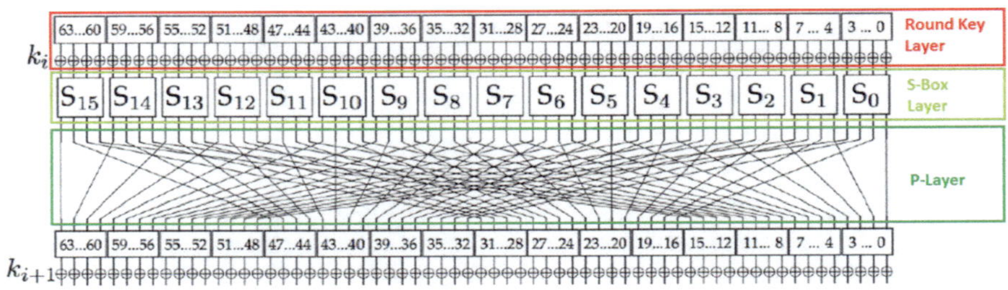

Fig. 3.9 The three layers at one round in the PRESENT cipher adapted from [7]

to the left and the round key is loaded with the 64 leftmost bits of the key register. Algorithm 1 outlines the PRESENT algorithm steps.

In the first layer at each round of addRoundkey, the STATE is updated by performing bitwise addition of the STATE with the round key. The second layer is sBoxLayer that consist of 16 copies of a 4-bit to 4-bit S-Box, S0-S15. The current state is divided into sixteen 4-bit words fed into S-Boxes. The content of the used S-Box in PRESENT is shown in Table 3.3. This is one of the smallest S-Boxes in hardware implementation with 28 GE for each.

The third layer is the pLayer, performing permutation on the bits of STATE. This layer changes the place of the bits in the STATE. Figure 3.9 shows one round of the PRESENT cipher composed of three layers.

PRESENT cipher uses S-Box and permutation components which are appropriate for FPGAs implementations. Implementing the PRESENT cipher on FPGA in [68] shows that only 117 LUT slices are need which makes it comparable in size to other ciphers.

The encryption and decryption processes in PRESENT cannot be processed in same hardware because the reverse of the S-Box is different in the decryption process from the encryption process. The implementation results of the PRESENT encryption cipher in different architectures is shown in Table 3.4 [54]. PRESENT also has been introduced to provide MAC with different key and output sizes.

Table 3.4 Hardware implementation results for PRESENT at 100 kHz frequency [54]

Design	Key size	Datapath width	Cycles/ block	Throughput [Kbps]	Tech. [μm]	Area [GE]	Current [μA]
Serialized	80	4	547	11.7	0.18	1075	1.4
Serialized	128	4	559	11.45	0.18	1391	N/A
Round-based	80	64	32	200	0.18	1570	2.78
Round-based	128	64	32	200	0.18	1884	3.67
Parallelized	80	64	1	6400	0.18	27,028	38.3

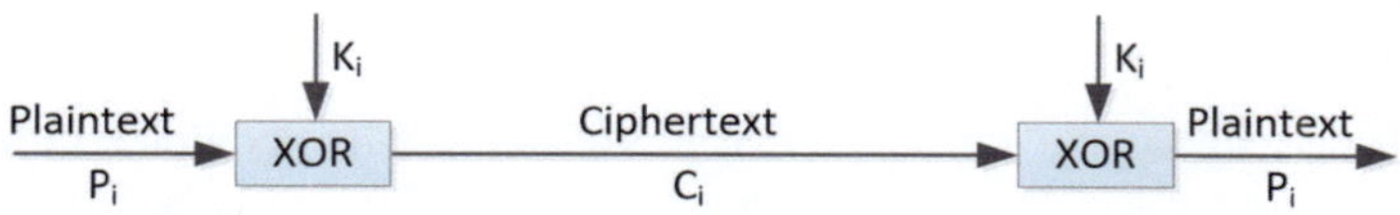

Fig. 3.10 One-time pad cipher

However, PRESENT is still away from the limitations of RFID systems because of its high area requirements.

PRESENT cipher has been proved that is secure against following attacks: statistical saturation [10], algebraic-differential attack, differential attack [67], linear with weak keys [51], multidimensional linear [9], bit-pattern integral [69], related key rectangle [52], linear hull [49]. However, it is shown in [42] that at the most 30 sub-key bits can be recovered by the attack given in [67] after some modifications in that algorithm. The authors of [63] have proposed improved side channel cube attacks which can reveal 48 bits of key with 211.92 chosen plaintext in PRESENT-80.

3.5.2 *Stream Ciphers*

A stream cipher is a function that processes the message bit by bit as a stream. They operate with a time-varying transformation on the individual plaintext digits inspired by the one-time pad concept.

One-Time Pad (OTP), also called Vernam-cipher [72], is a crypto algorithm wherein the plaintext is encrypted with a secret random key. The encryption is performed by having modulo 2 addition of the key and the message in a bit by bit manner. The decryption is accomplished by the same function (Fig. 3.10). If P_i, C_i and K_i are the plaintext, the ciphertext and the key bits, respectively, then:

$$\begin{aligned} &\textbf{Encryption}: \quad C_i = P_i \oplus K_i \qquad i = 1, 2, 3, \ldots \\ &\textbf{Decryption}: \quad P_i = C_i \oplus K_i \end{aligned} \tag{3.1}$$

It is proved that in OTP cryptosystem, the ciphertext will be impossible to decrypt or break unconditionally under the following conditions [35]:

- The key must be at least as long as the message or data that is being encrypted.
- The key must be truly random, not generated by a simple computer function.
- The key and plaintext are calculated by modulo 10 (digits), modulo 26 (letters) or modulo 2 (binary).
- Each key must be used only once, and both the sender and receiver must destroy their key after using it.
- There should only be two copies of the key: one for the sender and one for the receiver.

One-Time Pad has the advantage that it is faster and less complex in hardware than other cryptosystem but the challenge is the key which must be as long as the message. This challenge makes using it impractical for almost all applications. On the other hand, the decrypting party must have access to the same key to encrypt the message and this raises the problem of how to convey the key to the decrypting party safely, or how to keep both keys secure because of the difficulty in key distribution and management.

3.5.2.1 Keystream

Stream cipher is a practical scheme in which the infinite secret key of one-time pad cipher is replaced with a keystream. A keystream is a pseudorandom digit stream generated from a secret key of finite length while the keystream is independent of the plaintext and the ciphertext (Fig. 3.11). This scheme is close to one-time pad with the difference that the secret key is a seed to generate a stream for encryption and decryption.

The proposed scheme is theoretically never secure since the attacker can always try all possible 2^k keys as a brute force attack. Thus, the goal is to make it secure computationally. Since the keystream generator is only able to produce 2^k distinct keystreams, if a key is reused with a stream cipher in two different sessions, the exact same keystream will be produced. Hence, the attacker can easily find the keystream, and consequently the plaintext, by comparing two different ciphertexts.

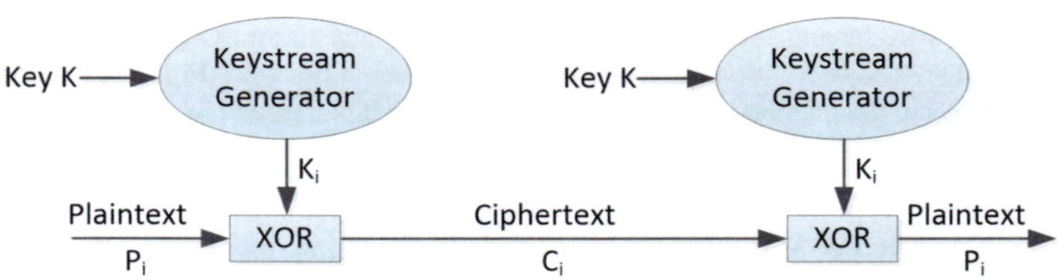

Fig. 3.11 Keystream generator scheme

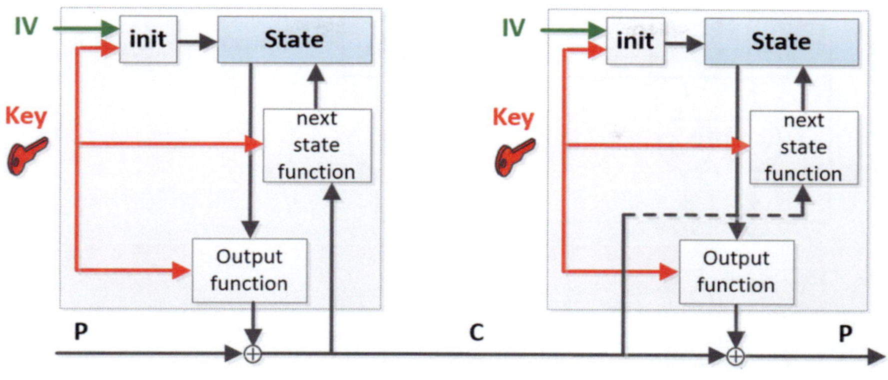

Fig. 3.12 Stream cipher operation

On the other hand, exchanging the key for each session is not practical. To solve this problem, modern stream ciphers utilize initial vector (IV). While the key is secret and constant between the sender and receiver, IV is public among all parties and after some sessions it is renewed and published over the network. The key and IV combined together will generate distinct keystreams for each session. Regarding Fig. 3.12, the operation of a stream cipher consists of the following phases:

1. In the initialization phase, the secret key and the public IV are loaded into a state register. The state is updated in some clock cycles, without producing any output to blend the key and the IV such that a change in the IV yields a completely different keystream. By setting up the internal state, the cipher will be ready for the next phase to generate the keystream.
2. In the encryption/decryption phase, the keystream is generated by updating the next state. Then the next block of data is encrypted/decrypted by the generated keystream.
3. After several communications, a new session starts by publishing another IV, while the secret key is same.

The main component of stream cipher is the keystream generator. This component is required to be capable of producing a long pseudo random sequence for any key while the security of the cipher does not depend on the IV. To build a keystream generator, there are some basic blocks and mathematical operations suitable for generating random streams such as linear feedback shift registers (LFSRs), with low complexity and good statistical properties, S-Boxes and Boolean functions to provide nonlinearity and bitwise addition (mod $2n$) which helps in making nonlinearity and breaking associativity.

A LFSR is a shift register whose present state is a linear function of its previous state as shown in Fig. 3.13. This register will produce a stream which will be repeated after a while. The length of the stream is dependent on its polynomial function $C(x)$ [Eq. (3.2)]. To have the maximum length, the LFSR function is required to be primitive.

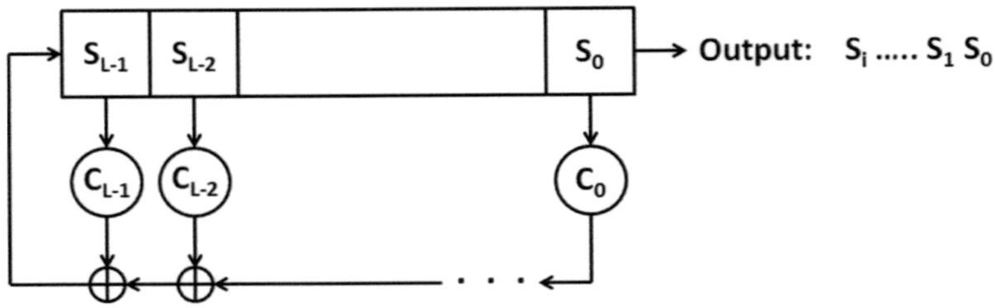

Fig. 3.13 Typical LFSR

$$C(x) = c_0 + c_1 x + c_2 x^2 + \ldots + c_{L-1} x^{L-1} \qquad c_i \in \{0, 1\} \tag{3.2}$$

A maximum length LFSR satisfies the pseudorandom assumption but the problem is the linear recursion. With 2^k output bits, the initial state can be recovered by using algorithms like Berlekamp-Massey [57]. To destroy nonlinearity, different stream ciphers have found different solutions. One of them is applying two or more outputs of the LFSR into a non-linear function. The other solution is applying a nonlinear filtering, e.g. a Boolean function and feeding it back to the input of LFSR which is called nonlinear shift register (NFSR). NFSR is a shift register which generates a nonlinear relation of the states. To provide non-linearity, a high linear function and a bent function will be used together. This approach is used in Grain and Trivium ciphers.

3.5.2.2 Trivium

Trivium is a synchronous stream cipher, designed to be compact in area and fast for high throughput applications [12]. Trivium supports 80-bit private key and 80-bit IV. It is composed of three NFSRs with different lengths of 93, 84 and 111 bits. In each clock cycle, the three NFSR registers are updated while the bitwise addition of their outputs generates the keystream (Fig. 3.14).

For initialization, the 80-bit IV is loaded into the first NFSR, the 80-bit key is loaded into the second NFSR while all the bits of the third NFSR are set to zero except the three last bits which are set to one. To start the encryption process, 1158 clock cycles are required before having the first output.

One of the advantages of Trivium is its small area. It can be implemented with 228 registers, 3 AND-gates and 7 3-input XOR-gates. The minimum area reported for implementing Trivium is 1294 GE In [28]. To speed up the cipher, it is possible to implement Trivium in parallel with different radix. Table 3.5 shows the results of two different implementations of Trivium with radix one.

Until 2010, no cryptanalytic attacks better than the brute force attack were known for Trivium. However, several attacks come close to it like the cube attack [16],

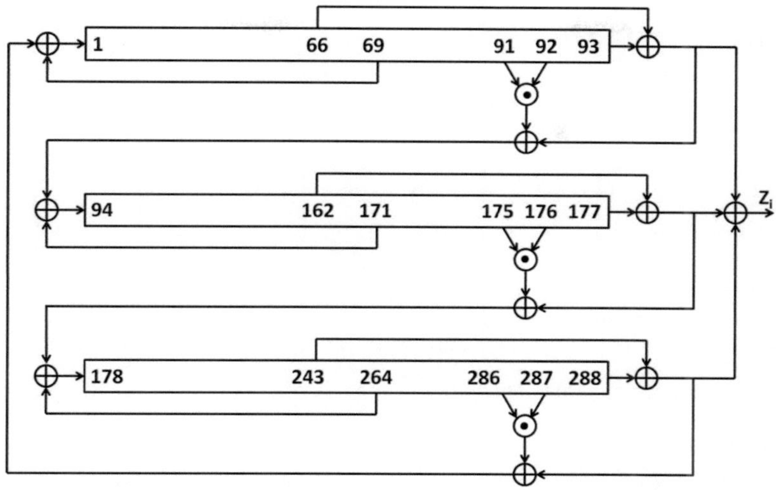

Fig. 3.14 Hardware implementation of Trivuim adapted from [12]

Table 3.5 Implementation results for Trivium

Design	Cycles init.	Tech. [μm]	Area [GE]	Power [μW] @ 100 kHz
Good and Benaissa [28]	1333	0.13	2599	5.6
Feldhofer [24]	1607	0.35	1603	1.06

the Algebraic IV Differential Attack (AIDA) [66], and also the proposed attacks in [47, 56] which made Trivium increase the length of the key beyond 80 bits.

3.5.2.3 Grain

Grain is a hardware-oriented stream cipher with small area overhead designed for limited resources environments. The first version of Grain supports 80-bit key and 64-bit IV [31]. The second version supports a key size of 128 bits and an IV size of 96 bits [2] along with optional authentication. The design of this cipher is very simple and based on two shift registers, one linear and one nonlinear, and three functions $f(x)$, $g(x)$ and $h(x)$ as shown in Fig. 3.15. $f(x)$ is a linear function while $g(x)$ and $h(x)$ are non-linear functions.

At the beginning of the encryption process, the LFSR and NFSR are initialized with the IV and key, respectively. Then the cipher is clocked 160 (first version) or 256 (second version) times without producing any keystream. The generated keystream is the output of $h(x)$. To speed up the Grain cipher, it is possible to implement it in parallel with different radix. Table 3.6 shows the results of two different implementations of Grain with radix one reported in 0.13 μm [28].

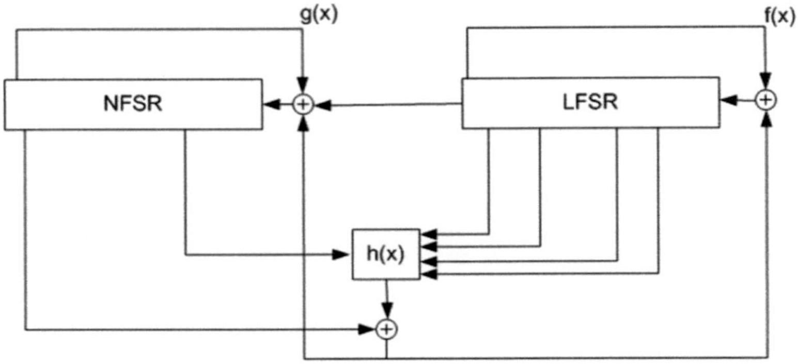

Fig. 3.15 Grain cipher adapted from [31]

Table 3.6 Implementation results for Grain cipher with different key sizes [28]

Key [bits]	Cycle init.	Cycle/ bits	Max freq. [MHz]	Area [GE]	Leakage power [μW]	Total power [μW] @ 10 MHz
80	321	1	724.6	1294	2.22	109.45
128	513	1	925.9	1857	2.70	167.73

The Grain cipher supports an optional authentication message with at least 32-bit size which will be appended to the end of the ciphertext before transmitting it. The implementation results of the Grain cipher in Table 3.6 does not cover the implementation of the authentication part.

To prevent substitution attacks in Grain, it is required to refresh the authentication key after each communication unless the key will be revealed after two or three communications [1]. The first version of Grain is vulnerable to a related key attack [40] and an algebraic attack with a weak Key-IV [70]. The second version of Grain is found to be immune against dynamic cube attacks and also differential attacks [45]. Until now, no attack is reported to break down Grain-128.

3.5.3 *Hybrid Ciphers*

Block ciphers and stream ciphers are two main groups of cryptosystems which are popular in lightweight cryptography. Each of these groups has its own advantages and disadvantages. Stream ciphers are interesting for two reasons. First, they are faster in software applications. This main advantage is important when an enormous amount of data is being encrypted like video streams. Another advantage of stream ciphers lies in their design perspective by having a low circuit complexity. However, stream ciphers require a considerable amount of time for initialization before generating the first output. This disadvantage is not important since initialization happens during the algorithm startup or whenever the key changes.

Compared with block ciphers which security is well understood particularly against statistical attacks, the security of stream ciphers still require more research. In block ciphers, several bits of data are encrypted together as a block. Therefore, some bits are required to be padded to the message when the message size is not a multiple of the block size. If any bits of the ciphertext change during transmission, it might be explored by the receiver in the encryption process since this bit may affect the entire message. Unlike block ciphers, each bit in stream ciphers is encrypted independently. Therefore, the sender and the receiver must be synchronized to confirm that the sender applies the right key sequence to the given bit of the plaintext. If any error happens during transmission, this error will be propagated to all bits of the ciphertext and all decrypted plaintexts will be wrong. Bit deletions or bit insertions are effective tools used by the attackers to break the cipher. Besides, the initial vectors may also bring new opportunities for the attackers.

Authentication in stream cipher is essential because the ciphertext is the result of the bitwise addition of plaintext and the keystream. Therefore, by changing one bit of the ciphertext, the corresponding change in the plaintext will be easily predictable. Meanwhile, the corresponding block is altered in a completely unpredictable way in block cipher. For example, if the plaintext contains an amount of money, the attacker might be able to alter this amount in stream ciphers by altering some bits of ciphertext.

Stream ciphers have the advantage that they encrypt and decrypt data with the same algorithm since the encryption process is the same as the decryption process. Meanwhile, a block cipher may need to have two different algorithms for the encryption and decryption processes which may impose extra area overhead for hardware implementation. On contrary, a block cipher is stateless such that the ciphertext is not a function of the time, while a stream cipher has an internal state.

Block ciphers and stream ciphers are compared in Table 3.7. Hybrid ciphers are those ciphers which inherit some of their properties from block ciphers and some other properties from stream ciphers to receive benefits from both approaches. The best example for these ciphers is the Hummingbird (HB) cipher.

Table 3.7 Comparing the properties of block ciphers and stream ciphers

Properties	Block ciphers	Block ciphers
Message size	Fixed (+padding)	Variable length
Memory	Stateless	Internal state
Core	Encryption + decryption	Encryption = decryption
Equivalent	Random permutation	PRNG
Model	Diffusion + confusion	One-time pad

3.5.3.1 Hummingbird (HB)

The Hummingbird (HB) cipher has a hybrid structure of block cipher and stream cipher, providing the designed security with block sizes as small as 16-bit block. It is specially designed for resource-constrained platforms. The first generation of HB, HB-1, was designed to provide 256-bit security and 80-bit internal state [20]. However, it was shown that HB-1 is vulnerable to a chosen-IV and chosen-message attacks in [62].

The Hummingbird-2 cipher has a 128-bit secret key and a 128-bit internal state which is initialized by a 64-bit Initialization Vector (IV) [21]. The used functions in HB-2 are the exclusive-or operation on words, addition modulo 65536 and a nonlinear mixing function $f(x)$. The nonlinear mixing function $f(x)$ consists of four-bit S-Box permutation lookups on each nibble of the word, followed by a linear mix. The fundamental block or round function of HB-2 encryption is defined as:

$$WD16(x, K_a, K_b, K_c, K_d) = f(f(f(f(x + K_a) + K_b) + K_c) + K_d) \tag{3.3}$$

where x is the input plaintext, intermediate state, K_a, K_b, K_c, and K_d are four 16-bit secret keys and the nonlinear function $f(x)$ is specified as:

$$\begin{aligned} S(x) &= S_1(x_1)|S_2(x_2)|S_3(x_3)|S_4(x_4); \qquad x = (x_1, x_2, x_3, x_4) \\ L(x) &= x + (x <<< 6) + (x <<< 10) \\ f(x) &= L(S(x)) \end{aligned} \tag{3.4}$$

The Hummingbird-2 S-Boxes S_1, S_2, S_3 and S_4 are given in Table 3.8.

For initialization, the intermediate states are loaded with IVs. Before starting the encryption process, four round procedures in 80 clock cycles will be run. The Hummingbird cipher is not reversible, therefore, for the decryption part different hardware is required to be implemented. The results of the hardware implementation of HB-2 with area and performance optimization are shown in Table 3.9.

Authentication is provided in HB-2 cipher and it is optional. To generate the authenticated message, no extra hardware is required and it can be generated by the same hardware used for encryption. The length of MAC is fixed to 64 bits for plaintext with sizes of one word to eight words.

Table 3.8 S-Boxes used in Hummingbird-2 [21]

x	0	1	2	3	4	5	6	7	8	9	10	11	12	13	14	15
$S_1(x)$	7	12	14	9	2	1	5	15	11	6	13	0	4	8	10	3
$S_2(x)$	4	10	1	6	8	15	7	12	3	0	14	13	5	9	11	2
$S_3(x)$	2	15	12	1	5	6	10	13	14	8	3	4	0	11	9	7
$S_4(x)$	15	4	5	8	9	7	2	1	10	3	0	14	6	12	13	11

Table 3.9 Hardware implementations of Hummingbird-2 [21]

Cipher	Tech. [μm]	Cycles / block	Datapath width	Area [GE]	Init. [cycles]	Throughput (bits/cycle)	Power [μW] @10 MHz
HB2-ee4c	0.13	4	16	3220	16	4	163.1
HB2-e16c	0.13	16	16	2332	80	1	156.8
HB2-e20c	0.13	20	16	2159	80	0.8	149.1

The security of HB-2 has been investigated in [8]. The authors have proposed an attack based on key recovery and differential sequence analysis (DSA) for HB-2. However, this attack is only of a theoretical interest and it does not affect the security of the Hummingbird-2 in practice. In [71], it has been proven that HB-2 cannot resist related key attacks.

3.6 Motivation for RBS Lightweight RFID Cryptosystems

Radio Frequency Identification is a pervasive technology offering more reliable, accurate and faster identification of objects compared to other solutions. Reading the data of the objects from distance regardless of bad weather or day-light in this technology makes it applicable in a wide range of applications. These advantages bring the hope that such technology will become the substitute of its old competitor, barcode, in the near future. However, to reach this goal there are still some obstacles which need to be considered and investigated.

Spreading RFID technology raises significant concerns in user privacy and security issues especially in secure applications such as authentication and payment systems. Having a secure communication is one of the main obstacles which will be removed by constructing a channel between the reader and the tag preserving confidentiality, integrity and authenticity. Confidentiality makes having access to data of objects difficult for attackers without the secret key. Integrity assures the received message is intact and authentication confirms the sender of the message in the channel.

Defending RFID devices against malicious attacks is possible only through a strong protection mechanism. Since physical solutions put limitations on tags and readers, they are suitable only for particular applications while providing privacy may be applicable partially. Cryptography is a respectable solution which can provide other services like integrity and authentication besides confidentiality. Contemporary cryptosystems are strong in terms of security. However, they require plenty of resources like power and area for their implementation while RFID tags are very resource constraint devices. Therefore, introducing new cryptosystems which can provide security with satisfying these limitations, called lightweight encryption is necessary for RFID systems.

Recently some lightweight encryption algorithms have been introduced for this purpose. However, these algorithms are more concerned with confidentiality while authentication is a part of privacy. On the other hand, current hash functions are not suitable for constrained environments. Since they require significant amounts of resources in their designs, they are not hardware friendly at all. In Part II of this book, a new symmetric encryption algorithm is presented which can provide confidentiality, integrity and authentication all together while the cost of its implementation is suitable for RFID systems.

3.6.1 RBS Design Objectives

The key design targets of our lightweight RFID cryptosystem to be presented in Part II of the book are summarized as follows:

- We aim at developing a new lightweight symmetric encryption cryptosystem—referred to as Redundant Bit Security (RBS). In this RBS cryptosystem, confidentiality of the plaintext is achieved through inserting some redundant bits inside the plaintext bits to change the location of plaintext bits inside the ciphertext. The location of the redundant bits and the plaintext bits inside the ciphertext is the secret key shared between the sender and receiver. Experimental results in Chap. 6 shows that the implementation of RBS requires less power and area compared to other known symmetric algorithms proposed for RFID systems especially when authentication is required. This saving in area overhead has a direct effect on the implementation cost of the RFID tags which is also one of the main concerns in getting acceptance by industry. Regarding other metrics such as the energy-per-bit, hardware efficiency, area-time product and power-area-time product, RBS offers better results. Since RFID tags are very resource constrained and they have strict limitations on the area and consuming power, the RBS algorithm is a promising candidate to provide confidentiality. In Chap. 5 its resilience against strong attacks is proved.
- Providing authentication and integrity: In RBS algorithm, the inserted redundant bits in the plaintext are calculated in such a way that they can provide authentication and integrity services along with confidentiality. Offering these services is very important in some applications like health care. Also in environments that the possibility of manipulating the transferred data is high by attackers or harsh condition. Implementing keyed hash function or MACs for this purpose requires high cost in area and power, while RBS offer these services with low cost.
- Flexibility in security level: The number of plaintext and redundant bits in the ciphertext are two important factors in defining the security level of the RBS cipher. By increasing each of them or both of them, the security level of the cipher will be increased. While increasing one of these parameters and decreasing the other one at the same time may lead to different results. Therefore, changing these two parameters gives the ability to the designer to change the security level

of the cipher to a desired level. The only part of hardware required to be updated with the security level is the MAC generator. If the number of redundant bits is constant, then the security level might be adjusted by the number of plaintext bits while the same MAC generator can be used without any change. For example, if there are 68 bits for redundant data, by changing the size of plaintext from 32 to 64 bits the key space will grow from 2^{86} to 2^{128}. Therefore, different keys can be supported with the same hardware. The only restriction is the size of redundant bits which should be longer than the plaintext at least for few bits to avoid collisions. However, if the number of redundant bits changes, the underling hardware is required to change.

3.7 Conclusion

In this chapter, the most recent and well-known symmetric and asymmetric ciphers designed for low-cost RFID implementation have been studied. These ciphers cover new lightweight designs like PRESENT, Gain and HB and also adapted and modified version of contemporary cryptosystems like ECC. Asymmetric ciphers provide key-management advantages and non-repudiation service besides confidentiality. However, these ciphers are computationally far more demanding than symmetric ciphers in terms of performance, power and area. This huge cost gap between these two types of ciphers makes asymmetric ciphers not suitable for RFID systems while new designs in cryptography are directed towards symmetric ciphers.

In symmetric algorithms, block ciphers and stream ciphers are both competitive candidates for obtaining the name of lightweight cryptography. Block ciphers are well investigated and understood in security, while stream ciphers are better in cost. The lightweight primitives presented in this chapter are further compared and discussed in Chap. 6, together with our proposed cipher presented in Chap. 4.

References

1. Agren, M., Hell, M., Johansson, T.: On hardware-oriented message authentication with applications towards RFID. In: Proceedings of International Workshop on Lightweight Security & Privacy [LightSec] (2011)
2. Agren, M., Hell, M., Johansson, T., Meier, W.: Grain-128a: a new version of grain-128 with optional authentication. Int. J. Wire. Mob. Comput. **5**(1), 48–59 (2011)
3. Batina, L., Mentens, N., Sakiyama, K., Preneel, B., Verbauwhede, I.: Public-key cryptography on the top of a needle. In: Proceedings of IEEE International Symposium on Circuits and Systems, ISCAS'07 (2007)
4. Bellare, M., Canetti, R., Krawczyk, H.: Keyed hash functions and message authentication. In: Advances in Cryptology - CRYPTO. Lecture Notes in Computer Science, pp. 1–15 (1996)
5. Biryukov, A., Canniere, C.D., Dellkrantz, G.: Cryptanalysis of SAFER++. In: Advances in Cryptology - CRYPTO 2003. Lecture Notes in Computer Science. Springer, Berlin (2003)

6. Bogdanov, A., Khovratovich, D., Rechberger, C.: Biclique cryptanalysis of the full AES. In: Advances in Cryptology - ASIACRYPT 2011. Lecture Notes in Computer Science, vol. 7073, pp. 344–371. Springer, Berlin (2011)
7. Bogdanov, A., Knudsen, L., Leander, G., Paar, C., Poschmann, A., Robshaw, M., Seurin, Y., Vikkelsoe, C.: PRESENT: an ultra-lightweight block cipher. In: Cryptographic Hardware and Embedded Systems - CHES 2007. Lecture Notes in Computer Science, vol. 4727, pp. 450–466. Springer, Berlin (2007)
8. Chai, Q., Gong, G.: A cryptanalysis of hummingbird-2: the differential sequence analysis. IACR Cryptology ePrint Archive (2012). http://eprint.iacr.org/2012/233
9. Cho, J.: Linear cryptanalysis of reduced-round present. In: Topics in Cryptology - CT-RSA 2010. Lecture Notes in Computer Science, vol. 5985, pp. 302–317. Springer, Berlin (2010)
10. Collard, B., Standaert, F.X.: A statistical saturation attack against the block cipher PRESENT. In: Proceedings OF CT-RSA 2009. Lecture Notes in Computer Science, vol. 5473. pp. 195–210. Springer, Berlin (2009)
11. Daemen, J., Rijmen, V.: The Design of Rijndael: AES - The Advanced Encryption Standard. Springer, Berlin (2002)
12. De Canniére, C.: Trivium: a stream cipher construction inspired by block cipher design principles. In: Information Security. Lecture Notes in Computer Science, vol. 4176, pp. 171–186. Springer, Berlin (2006)
13. Department of Commerce, U.S.: Data encryption standard. FIPS Publication (1977)
14. Department of Commerce, U.S.: Recommendation for the triple data encryption algorithm (TDEA) block cipher. Information Security (2004)
15. Diffie, W., Hellman, M.: New directions in cryptography. IEEE Trans. Inf. Theory **22**(6), 644–654 (1976)
16. Dinur, I., Shamir, A.: Cube attacks on tweakable black box polynomials. IACR Cryptology ePrint Archive (2008). http://eprint.iacr.org/2008/385
17. Eastlake, D.: US secure hash algorithm 1 (SHA1). RFC 3174 (2001)
18. Eberle, H., Gura, N., Shantz, S.C., Gupta, V., Rarick, L., Sundaram, S.: A public-key cryptographic processor for RSA and ECC. In: Proceedings of IEEE International Conference on Application-Specific Systems, Architectures and Processors (2004)
19. ElGamal, T.: A public key cryptosystem and a signature scheme based on discrete logarithms. IEEE Trans. Inf. Theory **31**, 469–472 (1985)
20. Engels, D., Fan, X., Gong, G., Hu, H., Smith, E.M.: Hummingbird: Ultra-lightweight cryptography for resource-constrained devices. In: Financial Cryptography and Data Security. Lecture Notes in Computer Science, vol. 6054, pp. 3–18. Springer, Berlin (2010)
21. Engels, D., Saarinen, M., Smith, E.: The Hummingbird-2 lightweight authenticated encryption algorithm. In: Proceedings of Workshop on RFID Security [RFIDSec] (2011)
22. Feldhofer, M., Rechberger, C.: A case against currently used hash functions in RFID protocols. In: On the Move to Meaningful Internet Systems 2006: OTM 2006 Workshops. Lecture Notes in Computer Science, vol. 4277, pp. 372–381. Springer, Berlin (2006)
23. Fan, X., Mandal, K., Gong, G.: WG-8: a lightweight stream cipher for resource-constrained smart devices. EAI Endorsed Trans. Secur. Saf. **15**(3), 151–157 (2015)
24. Feldhofer, M.: Comparison of low-power implementations of trivium and grain. eSTREAM, ECRYPT Stream Cipher Project (2007)
25. Feldhofer, M., Wolkerstorfer, J., Rijmen, V.: AES implementation on a grain of sand. In: IEE Proceedings - Information Security, vol. 152, pp. 13–20 (2005)
26. Gaubatz, G., Öztürk, E., Kaps, J.P., Sunar, B.: State of the art in ultra-low power public key cryptography for wireless sensor networks. In: Proceedings of IEEE International Conference on Pervasive Computing and Communications Workshops (2005)
27. Gilbert, H., Peyrin, T.: Super-Sbox cryptanalysis: improved attacks for AES-like permutations. IACR Cryptology ePrint Archive (2009). http://eprint.iacr.org/2009/531
28. Good, T., Benaissa, M.: Hardware results for selected stream cipher candidates. In: State of the Artof Stream Ciphers 2007 (SASC 2007), Workshop Record, pp. 191–204 (2007)

29. Hamalainen, P., Alho, T., Hannikainen, M., Hamalainen, T.D.: Design and implementation of low-area and low-power AES encryption hardware core. In: Proceedings of the 9th EUROMICRO Conference on Digital System Design, DSD '06 (2006)
30. Hankerson, D., Menezes, A.J., Vanstone, S.: Guide to Elliptic Curve Cryptography. Springer, Berlin (2004)
31. Hell, M., Johansson, T., Meier, W.: Grain - a stream cipher for constrained environments. Int. J. Wire. Mob. Comput. **2**(1), 86–93 (2007)
32. Hoffstein, J., Pipher, J., Silverman, J.: NTRU: a ring based public key cryptosystem. In: Proceedings of Algorithmic Number Theory (ANTS III) (1998)
33. Huang, Q., Kobayashi, H., Liu, B.: Modeling of distributed denial of service attacks in wireless networks. In: Proceedings of IEEE Pacific Rim Conference on Communications, Computers and Signal Processing (2003)
34. Juels, A., Weis, S.: Authenticating pervasive devices with human protocols. In: Advances in Cryptology - CRYPTO 2005. Lecture Notes in Computer Science, vol. 3621, pp. 293–308. Springer, Berlin (2005)
35. Kahn, D.: The Codebreakers. Macmillan, New York (1996)
36. Kaliski, B.: MD2 message digest algorithm. RFC 1319 (1992)
37. Kaps, J.P.: Cryptography for ultra-low power devices. Ph.D. thesis, Worcester Polytechnic Institute, Worcester, MA (2006)
38. Kaps, J.P., Sunar, B.: Energy comparison of AES and SHA-1 for ubiquitous computing. In: Emerging Directions in Embedded and Ubiquitous Computing. Lecture Notes in Computer Science, vol. 4097, pp. 372–381. Springer, Berlin (2006)
39. Kobilitz, N.: Elliptic curve cryptosystems. Math. Comput. **48**, 203–209 (1987)
40. Kucuk, O.: Slide resynchronization attack on the initialization of grain 1.0. eSTREAM, ECRYPT Stream Cipher Project (2006)
41. Kumar, S.S., Paar, C.: Are standards compliant elliptic curve cryptosystems feasible on RFID? In: Proceedings of Workshop on RFID Security (2006)
42. Kumar, M., Yadav, P., Kumari, M.: Flaws in differential cryptanalysis of reduced round PRESENT. IACR Cryptology ePrint Archive (2010). http://eprint.iacr.org/2010/407
43. Lee, Y.K., Sakiyama, K., Batina, L., Verbauwhede, I.: Elliptic-curve-based security processor for RFID. IEEE Trans. Comput. **57**(11), 1514–1527 (2008)
44. Lee, K.S., Chun, J.H., Kwon, K.W.: A low power CMOS compatible embedded EEPROM for passive RFID tag. Microelectron. J. **41**(10), 662–668 (2010)
45. Lehmann, M., Meier, W.: Conditional differential cryptanalysis of grain-128a. In: Cryptology and Network Security. Lecture Notes in Computer Science, vol. 7712, pp. 1–11. Springer, Berlin (2012)
46. Luo, P., Wang, X., Feng, J., Xu, Y.: Low-power hardware implementation of ECC processor suitable for low-cost RFID tags. In: Proceedings of Solid-State and Integrated-Circuit Technology (2008)
47. Maximov, A., Biryukov, A.: Two trivial attacks on trivium. IACR Cryptology ePrint Archive (2007). http://eprint.iacr.org/2007/021
48. Miller, V.: Uses of elliptic curves in cryptography. In: Advances in Cryptology - CRYPTO. Lecture Notes in Computer Science, pp. 417–426. Springer, Berlin (1985)
49. Nakahara Jorge, J., Sepehrdad, P., Zhang, B., Wang, M.: Linear (hull) and algebraic cryptanalysis of the block cipher PRESENT. In: Cryptology and Network Security. Lecture Notes in Computer Science, vol. 5888, pp. 58–75. Springer, Berlin (2009)
50. Nie, T., Zhang, T.: A study of DES and blowfish encryption algorithm. In: Proceedings of IEEE Region 10 Conference [TENCON] (2009)
51. Ohkuma, K.: Weak keys of reduced-round PRESENT for linear cryptanalysis. In: Selected Areas in Cryptography. Lecture Notes in Computer Science, vol. 5867, pp. 249–265. Springer, Berlin (2009)
52. Özen, O., Varıcı, K., Tezcan, C., Kocair, C.: Lightweight block ciphers revisited: cryptanalysis of reduced round present and hight. In: Information Security and Privacy. Lecture Notes in Computer Science, vol. 5594, pp. 90–107. Springer, Berlin (2009)

53. Öztürk, E., Sunar, B.: Low-power elliptic curve cryptography using scaled modular arithmetic. In: Proceedings of 6th International Workshop on Cryptographic Hardware in Embedded Systems (CHES). Lecture Notes in Computer Science, vol. 3156, pp. 92–106. Springer, Berlin (2004)
54. Poschmann, A.: Lightweight cryptography: cryptographic engineering for a pervasive world. Ph.D. thesis, Ruhr-University Bochum (2009)
55. Rabin, M.: Digital signatures and public-key functions as intractable as factorization. Technical Report mit/lcs/tr-212, Massachusetts Institute of Technology (1978)
56. Raddum, H.: Cryptanalytic results on trivium. eSTREAM submitted papers (2006). http://www.ecrypt.eu.org/stream/papersdir/2006/039.ps
57. Reeds, J., Sloane, N.: Shift-register synthesis (modulo m). SIAM J. Comput. **14**, 505–513 (1985)
58. Rijmen, V.: Practical-titled attack on AES-128 using chosen-text relations. IACR Cryptology ePrint Archive (2010). http://eprint.iacr.org/2010/337
59. Rivest, R.: The MD4 message digest algorithm. In: Advances in Cryptology - CRYPTO. Lecture Notes in Computer Science. Springer, Berlin (1990)
60. Rivest, R.: The md5 message-digest algorithm. RFC 1321 (1992)
61. Rivest, R., Shamir, A., Adleman, L.: A method for obtaining digital signatures and public-key cryptosystems. Commun. ACM **21**(2), 120–126 (1978)
62. Saarinen, M.J.O.: Cryptanalysis of Hummingbird-1. In: Fast Software Encryption. Lecture Notes in Computer Science, vol. 6733, pp. 328–341. Springer, Berlin (2011)
63. Shannon, C.E.: Communication theory of secrecy systems. Bell Syst. Techn. J. **28**, 656–715 (1949)
64. Sundaram, A.: An introduction to intrusion detection. Crossroads Magazine, Special issue on computer security. vol. 2(4). ACM, New York (1996)
65. Verbauwhede, I., Hoornaert, F., Vandewalle, J., Man, H.D.: Security and performance optimization of a new DES data encryption chip. IEEE J. Solid-State Circuits **32**, 647–656 (1988)
66. Vielhaber, M.: Breaking ONE.FIVIUM by AIDA an algebraic IV differential attack. IACR Cryptology ePrint Archive (2007). http://eprint.iacr.org/2007/413
67. Wang, M.: Differential cryptanalysis of reduced-round PRESENT. In: Progress in Cryptology - AFRICACRYPT 2008. Lecture Notes in Computer Science, vol. 5023, pp. 40–49. Springer, Berlin (2008)
68. Yalla, P., Kaps, J.: Lightweight cryptography for FPGAs. In: Proceedings of IEEE International Conference on ReConFigurable Computing and FPGAs, ReConFig '09 (2009)
69. Z'aba, M., Raddum, H., Henricksen, M., Dawson, E.: Bit-pattern based integral attack. In: Fast Software Encryption. Lecture Notes in Computer Science, vol. 5086, pp. 363–381. Springer, Berlin (2008)
70. Zhang, H., Wang, X.: Cryptanalysis of stream cipher grain family. IACR Cryptology ePrint Archive (2009). http://eprint.iacr.org/
71. Zhang, K., Ding, L., Gua, J.: Cryptanalysis of hummingbird-2. IACR Cryptology ePrint Archive (2012). http://eprint.iacr.org/2012/207
72. Zhao, X., Wang, T., Guo, S.: Improved side channel cube attacks on PRESENT. IACR Cryptology ePrint Archive (2011). http://eprint.iacr.org/2011/165

Part II
Lightweight RFID Redundant Bit Security

Chapter 4
RBS Cryptosystem

Abstract Redundant Bit Security (RBS) is a lightweight symmetric encryption algorithm that targets the resource-constrained RFID devices. Unlike its existing counterparts, RBS simultaneously provide confidentiality, authentication, and integrity of the plaintext by inserting hash-generated redundant bits among the already modified plaintext-data. Using a flexible-length hash algorithm in the RBS hardware implementation allows RBS to support different key sizes which implies flexibility in the security level. RBS hardware implementation consists of two parts. The first part implements the redundant bit generator by implementing a modified version of MAC generator that is compatible with block ciphers instead of the original MAC generator designed for stream ciphers. The second part of the RBS implementation targets implementing the encryption/decryption process. The encryption/decryption process is integrated with the transmission and reception modules of the RFID transponder. The RBS implementation presented supports online key size selection. This implies that the number of redundant bits, and accordingly the security level, in RBS is adjustable without changing the underlying hardware. This feature allows tags with the same hardware to operate with different key sizes.

Redundant Bit Security (RBS) is a lightweight authenticated symmetric encryption block cipher. The proposed RBS algorithm is light in terms of area and power consumption which makes it suitable for restricted-resource applications like RFID systems and sensor networks which are crucial members of the Internet of Things (IoT). Confidentiality of the plaintext in RBS algorithm is achieved by inserting some redundant bits inside the plaintext to change the location of plaintext bits while maintaining their order unchanged in the ciphertext. In addition to confidentiality, the redundant bits provide authentication and integrity services as well. The location of the redundant bits inside the ciphertext is a secret key shared between the two communicating parties. The security level of the RBS algorithm is adjustable through the choice of the number of redundant bits. The hardware implementation of RBS cipher requires less power and area compared to other known symmetric algorithms proposed for RFID systems which only provide confidentiality service.

Typically, encryption algorithms are based on performing complicated mathematical operations on the plaintext and ciphertext such as multiplications and

A. Khattab et al., *RFID Security*, Analog Circuits and Signal Processing,
DOI 10.1007/978-3-319-47545-5_4

divisions which consume plenty of resources. Unlike such conventional encryption methods, the RBS algorithm developed in this book does not rely on such complicated mathematics computations for the encryption and decryption processes. Instead, the message is intentionally manipulated by inserting redundant bits into the original bits. In the RBS algorithm, the location of the original bits changes in the ciphertext while their order is unaffected. As an example, suppose that the original message is "1010". Inserting one redundant bit at the third place changes the message to "10110". Knowing that just one bit of the ciphertext is redundant, the attacker confronts with five possibilities to find the place of the redundant bit. Besides the location of the redundant bits, their values are important in providing confidentiality as well. For instance, assume that the original bits are all zeros. Therefore adding one '0' bit as a redundant bit will not have the same effect as adding a "1" redundant bit in this case. Consequently, (1) the number of redundant bits, (2) their locations and (3) their values are all important factors in hiding the plaintext inside the ciphertext. In other words, there is a relation between the security level of the RBS algorithm and each of these parameters. In the following sections, the effect of such parameters on the security of the algorithm and the way they are calculated are investigated. At the end of this chapter, the hardware implementation of RBS algorithm will be presented.

4.1 Key and Number of Redundant Bits

The first parameter of the RBS algorithm is the number of redundant bits required to provide a desired security level. Once the number of redundant bits is calculated, the hardware is built accordingly. Then all of the communications between the two communicating parties will be encrypted using this number of redundant bits. Hence, the number of redundant bits is a public parameter in this algorithm which is published to all parties. Meanwhile, the location of these redundant bits inside the key is secret between only the sender and the receiver.

The security level of the RBS algorithm is proportional to the number of redundant bits. Increasing the number of redundant bits increases the number of possible plaintexts a potential adversary will deal with, which makes finding such redundant bits in a ciphertext more complicated. On the other hand, the security level is defined by the size of the key space. This definition relates the size of the key space and the number of redundant bits.

4.1.1 Key Space

The security level is defined by answering the question of how long it will take an attacker to break the algorithm based on what resources the attacker needs in order to have a reasonable chance of succeeding. The cost of breaking the algorithm, usually measured in time and/or money, is required to be higher than the value of the protected asset. For example, if the information is needed to be secure for only a few hours, 1 week effort for breaking of the system might be acceptable. One of the well-known tools for measuring the security level of an algorithm is the key space.

The key space determines the set of all possible keys that can be used to initialize a cryptographic algorithm. The security level of an encryption algorithm has a direct relationship with its key space. Suppose that n is the number of original bits or plaintext and m is the number of redundant bits. The ciphertext is an $(n+m)$-bit data obtained by inserting the redundant bits among the plaintext bits. The location of redundant bits inside the ciphertext defines the secret key. Therefore, the secret key is simply an $(n + m)$-bit string wherein a "1" represents the location of a redundant bit and a "0" represents the location of plaintext bit in the ciphertext. For example, suppose that "10", "01" and "0110" are plaintext, redundant data and secret key respectively. The first and last bits of the ciphertext belong to the plaintext and the other bits of the ciphertext will be replaced by redundant bits. For the sake of simplicity, suppose that the plaintext data appears directly in the ciphertext without any alteration. Under this assumption, the ciphertext would be "1010". However, in RBS algorithm the plaintext bits are altered before inserting them to the ciphertext which will be discussed later.

The number of possible keys, i.e. the key space, is computed by counting all the possibilities to select m distinct elements from a set of $n + m$ elements, which is referred to as m-combinations. The size of key space (s) or the number of possible locations of the redundant bits in the ciphertext, depends on n and m and is expressed as

$$s = \binom{m+n}{n} = \frac{(m+n)!}{m!n!} = \frac{\prod_{i=1}^{m}(i+n)}{\prod_{i=1}^{m} i} \tag{4.1}$$

In Eq. (4.1), m and n are interchangeable. In other words, increasing either the number of redundant bits or the number of plaintext bits has the same effect on the key space size. As a result, the size of key space can be adjusted to the desired security level by fixing one of the two parameters and changing the other parameter. This is different from increasing one of these two parameters while decreasing the other parameter.

Figure 4.1 illustrates how the key space grows by changing the number of redundant bits and plaintext when the total number of bits is constant ($m+n = 100$). Figure 4.1 shows that when these two parameters are far from each other, the key space will reach its minimum size. The maximum size of key space happens when the number of redundant bits is equal to the size of the plaintext. Based on this graph, it can be concluded that high security levels for a small block of plaintexts is not possible unless a huge number of redundant bits is used. Furthermore, it is not possible to provide a high security level with a small number of redundant bits. The best choice is selecting these two parameters close to each other in order to obtain high security levels.

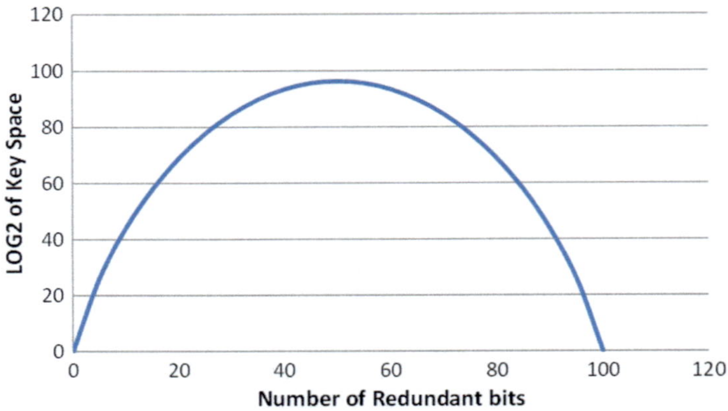

Fig. 4.1 Changing the size of the key space with the number of redundant bits

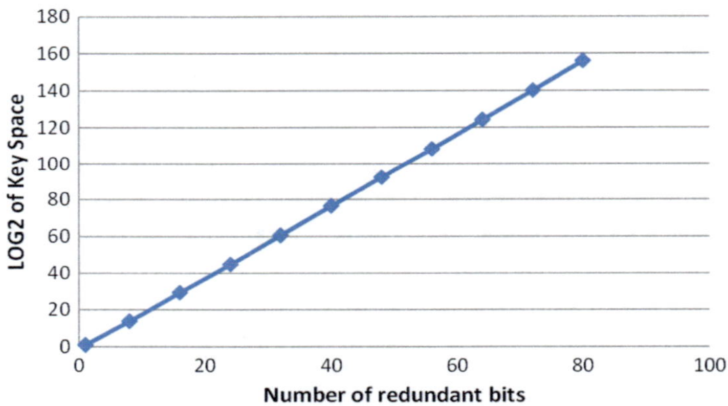

Fig. 4.2 The size of the key space when the number of redundant bits is equal to the plaintext bits

To find the optimum number of redundant bits, the size of the key space is calculated when the number of redundant bits and the size of the plaintext are equal. It is the situation in which the key space is in its maximum size. Figure 4.2 illustrates how big the key space can be for different sizes of plaintexts when the same size of redundant bits is merged with it.

The starting point of this study is selected to be a 64-bit block plaintext. Figure 4.3 demonstrates the relationship between s and m for $n = 64$. Increasing m from 0 to 128, the key space (s) will exponentially increase from 1 to 2^{172}.

As mentioned earlier, there is a relationship between the size of the key space of an encryption algorithm and its security level against possible attacks. The question is how big the key space should be to guarantee the desired security level. The Brute Force attack has been studied for finding the boundary of the key space for the RBS algorithm. In this attack, the attacker performs a complete search through all

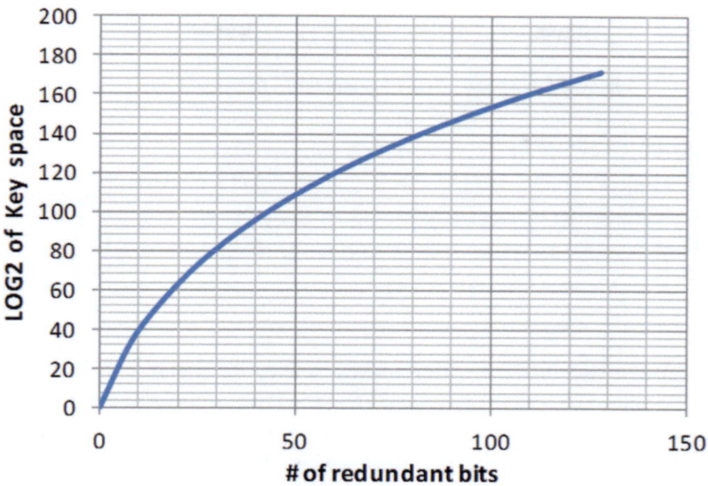

Fig. 4.3 The growth of the key space while the plaintext size is fixed to 64 bits

Table 4.1 The number of bits required in the ciphertext to have $s = 2^{128}$

m	50	55	57	60	63	64	65	66	67	68	70	73	78	91
n	91	81	78	73	70	68	67	66	65	64	63	60	57	50
c^*	140	136	135	133	133	132	132	132	132	132	133	133	135	140

c^*: Number of bits in the ciphertext

of the possible keys in the key space to find the right key. The 2^{128} key space size is computationally secure against the Brute Force attack. Applying this number in Eq. (4.1), there will be a variety of choices for m and n.

Table 4.1 shows a possible set of m and n for $s = 2^{128}$. One of the limiting factors which makes the span of choices narrow is the size of the ciphertext. Considering the fact that the required energy for transmitting the message increases with the length of the ciphertext, m and n should be chosen such that the number of bits of the ciphertext becomes the least. Referring to Table 4.1, the minimum size of the ciphertext is 132 bits and this happens when (n, m) are equal to any of the values (64,68), (65,67), (66,66), (67,65), or (68,64) which are highlighted in Table 4.1.

The best choice is (64, 68), i.e. when the size of the plaintext is 64 bits with 68 bits redundant bits, for two reasons. First, data blocks are processed and stored normally in multiples of 8-bits. While the redundant bits are used only in this cipher so they do not need to be a multiple of eight. Second, the number of redundant bits is required to be more than the plaintext bits by few bits to prevent collisions as will be discussed later in this chapter.

Based on the application requirement, the strength of security may change. The designer can change the number of plaintext and redundant bits to reach a desired security level. Performing the same simulation steps for $s = 2^{128}$, the recommended

Table 4.2 The number of required redundant bits for different security levels

	Size of key space	Size of plaintext	Number of redundant bits	Size of key & ciphertext
RBS-83	2^{80}	40	43	83
RBS-100	2^{96}	48	52	100
RBS-116	2^{112}	56	60	116
RBS-132	2^{128}	64	68	132
RBS-197	2^{192}	96	101	197
RBS-262	2^{256}	128	133	261

number of redundant bits for different key spaces has been acquired as shown in Table 4.2. Compared to other cryptosystems, the RBS algorithm needs extra 3–4 bits in the key size to provide the same security level that other cryptosystems support.

In Table 4.2, the size of the plaintext varies between 40 and 64 bits to provide key spaces between 2^{80} and 2^{128}. However, the designer has the ability to change the number of plaintext and redundant bits to reach any arbitrary security level. For example, to have 2^{80} key space when the plaintext size is 32 bits, the designer can either use 56 redundant bits or 43 redundant bits as the case in RBS-83. However, the first choice results in 88 bits in the ciphertext which is 5 bits longer than the ciphertext in RBS-83.

Using plaintexts shorter than 40 bits reduces the security level of RBS cipher sharply unless the number of redundant bits grows dramatically. Using plaintexts longer than 64 bits increases the security level more than necessary. In both cases, the length of the ciphertext becomes very long for transmission. On the contrary, 2^{128} key space can be obtained by 128-bit plaintext along with 40-bit redundant. Thus, the length of the ciphertext will be shorter than when RBS-132 is used for two 64-bit plaintexts. However, this design is not acceptable since the number of redundant bits is less than the plaintext and it cannot guarantee exclusive redundant data for each plaintext. On the other hand, the hardware overhead will be significantly high for this design. Thus, it is recommended that the size of the plaintext is limited between 40 and 64 bits based on the desired security level. For plaintext shorter than 40 bits, RBS-83 is a good choice. For plaintext longer than 64 bits, the plaintext will be broken down into suitable sizes.

4.1.2 Flexibility in Security Level

Supporting different key sizes with the same hardware is one of the advantages of RBS algorithm. The number of plaintext and redundant bits in the ciphertext are two important parameters in defining the security level of RBS. Tuning these two parameters gives the ability to the designer to change the security level of the cipher to a desired level. Before hardware implementation, the optimum cipher required for the given key size can be designed off-line. Such a flexibility may not be easy for other block ciphers which use pre-defined key and data block sizes.

After implementation, the designer still has the ability to change the security level of RBS cipher online by using different key and data block sizes while the number of redundant bits in the key is same as before. The only part of RBS hardware which is fixed and cannot be updated with the security level is the MAC generator. Therefore, the number of redundant bits is constant in different key sizes and the security level can be adjusted only by the number of plaintext bits while the same MAC generator can be used without any changes. For example, if there are 68 bits for redundant data, by changing the size of the plaintext from 32 to 64 bits, the key space can vary from 2^{86} to 2^{128}. Therefore, different tags with different key sizes can be supported with the same hardware. By using this feature, it will not be required to replace the tags whenever the security level of the system changes. The only restriction in key flexibility is the size of redundant bits which should be longer than the plaintext at least with few bits to avoid collisions.

4.2 Location of Redundant Bits

The second significant parameter in achieving the security of RBS algorithm is the location of the redundant bits inside the ciphertext. This information is to be kept as a secret key among the involved parties in order to have a secure communication. Revealing the location of any redundant bits in the ciphertext will diminish the size of key space.

The distribution of redundant bits inside the ciphertext is also another important factor in providing confidentiality. This distribution should not follow any linear or non-linear mathematic function, otherwise (1) the size of the key space will be reduced, (2) a dependency among the redundant bits will be constructed, and (3) the redundant bits will be distributed uniformly among plaintext bits. Therefore, the position of every redundant bit must be independent of other bits' positions. This way, if the location of one of the redundant bits being exposed, just the key space will shrink while the location of other redundant bits is still secret.

The best solution is utilizing a random distribution of the redundant bits inside the ciphertext. Such a distribution can be defined by the user or by using a pseudo random number generator (PRNG) with the condition that the number of ones and zeros in the secret key are constant.

4.3 Value of Redundant Bits

In addition to providing confidentiality of the sent data, the injected redundant bits can carry additional information about the original data as well. In order to generate these redundant bits, there are three options:

- **Choosing constant values for the redundant bits.** In this case, the redundant bits are the same for different plaintexts. This way the attacker can easily figure out the location of the redundant bits just by comparing the ciphertexts of two or more different plaintexts.
- **Choosing random values for the redundant bits.** In this case, there would be several ciphertexts for one plaintext. This way the attacker can also easily figure out the location of the redundant bits by comparing the different generated ciphertexts for the same plaintext.
- **The values of the redundant bits are injective functions of the plaintext.** Hence, there is an exclusive redundant data per each plaintext. Consequently, the plaintext and redundant data cannot be distinguished easily in the ciphertext.

Among these three approaches, the third option is the most suitable as it has the potential to provide both attack prevention and authentication. This algorithm can be implemented by splitting the plaintext into small blocks and performing mathematical functions on each block individually. At the end, all blocks are combined and then encrypted by a secret key. The pseudo code of such a redundant data generation algorithm is presented in Algorithm 2.

One applicable implementation solution for the algorithm presented in Algorithm 2 is through a Message Authentication Code (MAC) algorithm because a very small change in the plaintext will produce an entirely different output. Using MAC algorithm for generating the redundant bits, integrity and authentication will be provided as well as confidentiality which will be discussed in the following subsection.

4.3.1 Message Authentication and Data Integrity

Data integrity is defined as maintaining the correctness and consistency of a message. Since the message is sent via a wireless network, the integrity of the message is always in danger of being altered in transmission by either an adversary or environmental hazards, such as heat, dust, and electrical surges. Therefore, the receiver should validate the received data.

Algorithm 2 Redundant data generation algorithm

Split the plaintext into several small segments S_i
for each i **do**
    Shift/rotate/add/XOR (S_i, a constant number N_i)
end for
Combine all segments S_i to a single segment S
Encrypt (S, secret key K) using a symmetric algorithm

//K will be used in the receiver side for authenticating the sender

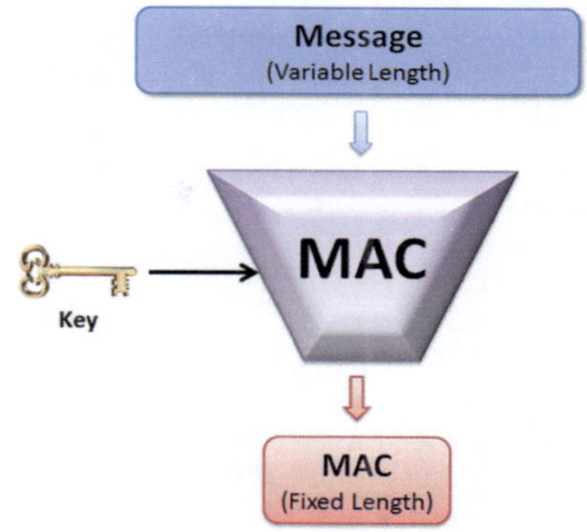

Fig. 4.4 MAC algorithm block diagram

Message authentication is one of the cryptography services which guarantees that the received message has been sent by an eligible user. It is crucial for a party—tag or reader—which receives a message to make sure who sent it. The Message Authentication Code or MAC is a piece of information which is used for both data integrity and authentication purposes. It is generated by a MAC algorithm which has two inputs (1) an arbitrary-length message, and (2) a shared key between two parties (Fig. 4.4). Typical MAC algorithms are strong in terms of security and some of them also guarantee that no collision will happen in their outputs for different input messages.

In general, there are three existing protocols for embedding the MAC inside the ciphertext (Fig. 4.5).

- The first protocol is more common than other protocols in symmetric key encryption algorithms. In this protocol, the MAC is first generated using the authentication key, K_{mac}. Then, the MAC will be attached to the original message and then the new message is encrypted by an encryption key, K_{enc} (Fig. 4.5a) [6]. After decrypting the received data at the receiver side, the MAC will be regenerated and then compared with the received one. Generating the same MAC means that the message is intact and it is sent by an authorized party. Otherwise, the received message will be discarded.
- The generated MAC will be generated and attached to the end of the ciphertext before transmission (Fig. 4.5b). Grain [2] and HB-2 [5] both use this protocol for providing authentication.
- Instead of the plaintext, the MAC of the encrypted plaintext is attached to the encrypted message before transmission (Fig. 4.5c). This protocol requires more time than other two protocols since the MAC generation and encryption cannot happen at the same time.

In the second and third protocols, the boundary between the MAC and message is clear. Hence, the MAC algorithms used in these protocols are required to be very secure against the substitution attack. In this attack, the adversary tries to replace the legitimate message with its own plaintext and MAC and assumes that it will

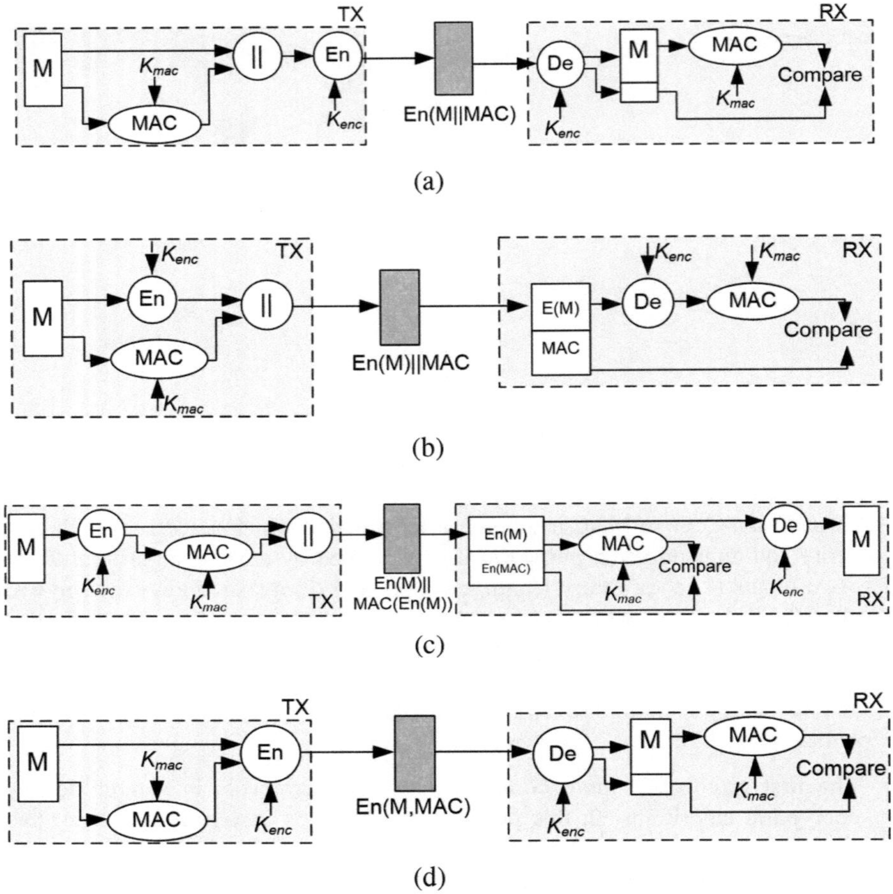

Fig. 4.5 Embedding the MAC inside the ciphertext in different existing protocols. (**a**) First authentication protocol. (**b**) Second authentication protocol. (**c**) Third authentication protocol. (**d**) Proposed authentication protocol

be accepted by the receiver. Considering this fact, the first protocol is more secure against substitution attack because the MAC is encrypted along with the plaintext and there is no direct access to it.

4.3.2 Message Authentication and Redundant Bits

As stated earlier, the redundant data is generated by the MAC algorithm. The second and third MAC generation protocols cannot be used in the RBS algorithm as the MAC is attached to the end of the message which is the redundant part in RBS.

The method used in RBS is based on a modified version of the first protocol (Fig. 4.5d). It might be noticed that second and third protocols can be special cases of RBS algorithm when all of the redundant bits are located at the end of the ciphertext. In other words, if m is the number of redundant bits, the m most significant bits of the secret key are ones while the rest of the key bits are zeros.

In the RBS algorithm, the generated MAC as redundant bits is inserted among the message bits instead of being appended to the end of the message. In other words, merging the MAC with the plaintext is a part of the encryption process. The distribution pattern of the MAC part inside the ciphertext is based on the encryption key. At the receiver side, the received data is broken into two parts based on the encryption key: the altered plaintext and the redundant bits. The receiver decides whether to keep the data or discard it by regenerating the MAC of the received plaintext at the receiver side and comparing it with the received MAC.

4.4 Plaintext Manipulation

How the plaintext appears inside the ciphertext is the last significant parameter in RBS algorithm which is directly related to the confidentiality of the algorithm. Three possible scenarios exist for such a task that will be discussed in the following subsections.

4.4.1 Direct Appearance Inside the Ciphertext

In this approach, the original plaintext bits will be merged with the redundant bits without any change in the plaintext. Therefore, the plaintext can be easily extracted from the ciphertext in the decryption process by removing the redundant bits. Despite the simplicity of this method, the key space may shrink sharply which makes the algorithm vulnerable to some attacks such as the known plaintext attack and the chosen plaintext attack. In these attacks, the attacker knows the plaintext. Hence, those bits of the ciphertext which have the same value of the plaintext will be potential locations for plaintext bits in the secret key. For example, if the plaintext is all zeros, all the corresponding zeros in the ciphertext might be locations of the plaintext in the secret key too.

There are some ways to expand the key space size such as increasing the number of redundant bits or having separate encryption keys based on the plaintext pattern. Increasing the number of redundant bits introduces more area overhead in the MAC implementation, and consequently, more power for transmitting the ciphertext. Generating a new key based on the plaintext pattern and exchanging it are also challenging tasks in symmetric encryption algorithms. Based on the stated reasons, this approach is not appropriate to be used in the RBS algorithm.

4.4.2 Bitwise Addition with a Constant-Value Keystream

In this approach, regardless of the pattern of the plaintext, some bits of the plaintext will always be altered in the ciphertext. The location of these plaintext bits is fixed. Thus, always some bits of the plaintext will appear altered while other bits are not changed. This approach makes the algorithm secure against known plaintext attacks because the attacker does not know which bits of the plaintext are altered in the ciphertext. As a solution for this approach, suppose that the secret key, K_{enc} is a binary sequence such that $K_{enc} = \{k_0, k_1, \ldots, k_{n+m}\}$ where n is the size of plaintext and m is the number of redundant bits. This key will be broken into two binary sequences K_{enc1} and K_{enc2} such that:

$$\begin{aligned} K_{enc1} &= \{k_0, k_1, \ldots, k_i\} \\ K_{enc2} &= \{k_{i+1}, k_{i+2}, \ldots, k_{n+m}\} \end{aligned} \tag{4.2}$$

where $i = \frac{n+m}{2}$. Then, the plaintext will be XOR-ed by $K'_{enc} = K_{enc1} \bigoplus K_{enc2}$. This way, some bits of the plaintext will be altered depending on the value of K_{enc}. The number of altered bits varies between zero bits (when $K_{enc1} = K_{enc2}$) to n bits (when $K_{enc1} =\sim K_{enc2}$). Since the attacker does not know the encryption key, it does not know how many bits of the plaintext and which of them in the ciphertext have been altered. This way, the attacker confronts with 2^m different possible manipulated plaintexts. However, this solution is still vulnerable to chosen plaintext attacks. Suppose that, the attacker changes just one bit of the plaintext, the redundant bits (MAC) in the ciphertext will change while all the bits of the plaintext but one, will be same. Comparing these two ciphertexts which have almost the same plaintexts will shrink the key space dramatically and make it easy for the attacker to find the approximate location of the plaintext in the ciphertext. Using this approach is not promising due to its weakness against such attacks.

4.4.3 Bitwise Addition with Variable-Value Keystream

In this approach, the plaintext bits are altered by performing bitwise addition of the plaintext with a variable keystream. The keystream will be different for each plaintext and it is unique for each plaintext such that for the same plaintexts, the same ciphertexts will be generated. To satisfy this condition, the keystream is required to be a function of the plaintext. This way, if any change happens in the plaintext, different keystreams and consequently different ciphertexts will be produced.

This solution is somehow similar to one-time pad because in both methods the plaintext is altered with a variable keystream. However, it is different since in one-time pad the keystream is a randomly generated number. Therefore, for two similar plaintexts, different keystreams will be generated and the keystream is independent of the plaintext. Meanwhile, there is a dependency between the plaintext and the generated keystream in the proposed solution.

Varying the value of the keystream based on the ciphertext will eliminate the weaknesses witnessed in the last approaches since for each plaintext the number of altered bits and their locations will be different from any other plaintext. This feature encourages us to use this approach for manipulating the plaintext in the ciphertext.

One straightforward mean for implementing this approach is through the MAC function. For the sake of hardware efficiency, the same hardware for generating the redundant data can be used for altering the plaintext. However, the MAC of the plaintext has already been used as redundant data, and hence, it cannot be used again as a keystream. Otherwise, it creates dependency between the altered plaintext and the redundant data which makes the algorithm vulnerable to some attacks such as the chosen plaintext attack. Instead of using the MAC(Plaintext) as a keystream, the MAC(Redundant data), or more precisely MAC(MAC(Plaintext)), is used for generating the keystream in the RBS algorithm as illustrated in Fig. 4.6a.

As Fig. 4.6a shows, the altered plaintext is obtained by performing bitwise addition of the plaintext with the generated keystream. Eventually, the ciphertext will be produced by merging the altered plaintext with the redundant data based on the secret key. Since the length of the keystream is not equal to the length of the redundant data—it is three to four bits longer than the length of the plaintext data—only the n least significant bits of the keystream (n is the size of the plaintext) will be used to generate the altered plaintext. The decryption process is illustrated in Fig. 4.6b. The receiver side extracts the redundant part from the ciphertext with the secret key. Afterwards, the keystream will be generated through the MAC of the redundant data which then will be used for recovering the original plaintext.

Having the same process for encryption and decryption is one of the main advantages of the RBS algorithm which makes the same hardware implementation usable for both processes. This characteristic, which has already been studied in stream ciphers, will cause significant saving in the area.

4.5 Implementation

The hardware implementation of RBS is composed of three main modules: the MAC generator, the encryption module, and the decryption module which will be discussed in detail in the following subsections.

4.5.1 MAC Generator

The MAC generator circuit is responsible for generating the MAC of the plaintext utilized as redundant bits in the ciphertext and also MAC(MAC(P)) or the MAC of the redundant bits used as the keystream. Table 4.3 summarized the main feature of existing MAC generators.

Fig. 4.6 Block diagram of encryption and decryption. (**a**) RBS encryption. (**b**) RBS decryption

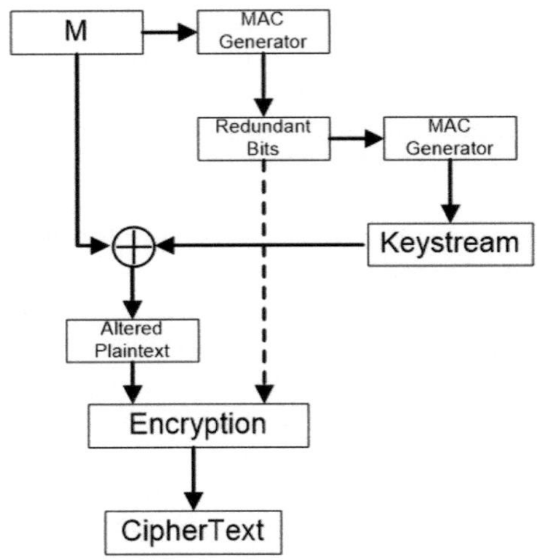

(a)

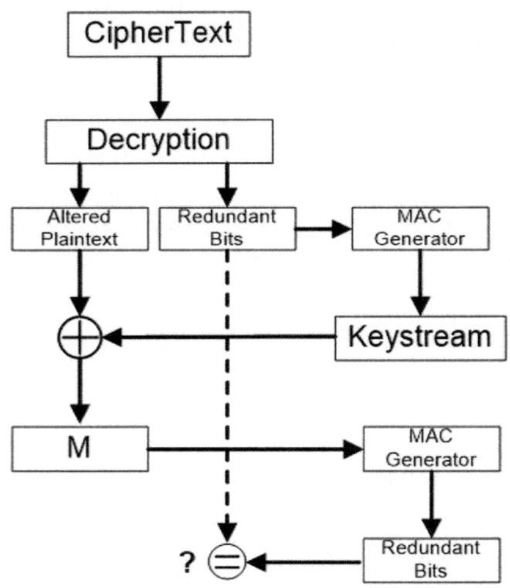

(b)

The implementation complexity of the MAC generator depends on its underlying MAC algorithm. There are several categories of MAC algorithms. A number of these algorithms uses block cipher operations such as OMAC, CBC MAC, and PMAC. The CBC-MAC comes in different versions that vary in details such as padding, length variability and key search strengthening [3].

HMAC is another type of MACs which is based on iterating a hash function. The cryptographic strength of HMAC depends on the cryptographic strength of the underlying hash function, the size of its hash output, and the size of the key. The size of the output of HMAC is equal to the size of the output of the used underlying hash function.

Table 4.3 Summary of MAC algorithms

Algorithm	Output size	Comments
CBC-MAC	Varies in versions	Based on block ciphers
HMAC	Fixed depends on used hash function	Based on hash functions
VMAC	Multiple of 64 bits	Based on block cipher and universal hash, good in software
UMAC	Support different sizes	Based on universal hash and choosing hash function randomly

Implementing MACs by universal hash functions has been popular because it was shown to be secure and well adapted for hardware implementation [8]. VMAC is a block cipher-based MAC algorithm which is using a universal hash. VMAC has excellent performance in software implementation. The length of VMAC output is a multiple of 64 bits up to the block size of the block cipher in use. UMAC is a type of MACs based on universal hashing which is calculated by choosing a hash function from a class of hash functions.

One of the main factors in selecting a MAC algorithm in RBS is the length of the generated MAC which is usually fixed and is defined as a parameter of the algorithm. For instance, SHA-0 and MD5 generate 160-bit and 128-bit MACs, respectively. In other words, the size of MAC depends on the chosen MAC algorithm. This limitation can be resolved if the chosen MAC algorithm supports a variable-length MAC.

4.5.2 Chosen MAC Algorithm for RBS

The MAC algorithm presented in [1] is a family of universal hash functions. This MAC is selected to be utilized in the RBS algorithm for two reasons. First, the output of this MAC can be set before implementation to generate arbitrary size MACs. Second, it is a lightweight universal hash algorithm from ϵ-almost XOR universal (ϵ-AXU) family based on Toeplitz matrices. The security of this algorithm is promised by low probability of exact substitution. Moreover, its resistance against collision attacks is high.

Let H be a family set of hash functions mapping from set A to set B by (H, A, B). Then (H, A, B) is defined to be ϵ-AXU if $\forall x, x' \in A, x \neq x', y \in B$,

$$| \{h \in H : h(x) \oplus h(x') = y\} | \leq \epsilon \mid H \mid \tag{4.3}$$

Constructing a MAC using an ϵ-AXU family, one part of the key is used to select a function from $h \in H$ and the output of this function is XOR-ed with a second part of the key, used as a one-time pad, chosen randomly from B. The MAC algorithm proposed in [1] is constructed based on Toeplitz matrices by assuming that k_i, $i = -w, 1-w, \ldots, L-2$ is a sequence of randomly chosen key bits. Then if $\mathbf{t} = (t_0, \ldots, t_{w-1})$ is a bit-vector of tag or digest with length w, and $\mathbf{m} = (m_0, \ldots, m_{L-1})$ is a bit-vector of message with length L, a possible MAC construction will be:

$$\begin{bmatrix} t_0 \\ t_1 \\ \vdots \\ t_{w-1} \end{bmatrix} = \begin{bmatrix} k_{-1} & \ldots & k_{L-2} \\ k_{-2} & \ldots & k_{L-3} \\ \vdots & \ddots & \vdots \\ k_{-w} & \ldots & k_{1-L-w} \end{bmatrix} \begin{bmatrix} m_0 \\ m_1 \\ \vdots \\ m_{w-1} \end{bmatrix} \tag{4.4}$$

From algorithmic interpretation point of view, the output of MAC or tag is initialized as $\mathbf{t} \leftarrow 0$. Introducing a window of size w to form:

$$\begin{aligned} \mathbf{K}_0 &= [k_{-w} \quad \ldots \quad k_{-1}] \\ &\vdots \\ \mathbf{K}_{L-1} &= [k_{L-1-w} \quad \ldots \quad k_{L-2}] \end{aligned} \tag{4.5}$$

For each bit m_i, nothing will happen if the bit is zero. If m_i is one, the tag will be updated by $\mathbf{t} \leftarrow \mathbf{t} \oplus K_i$. Figure 4.7 shows the block diagram of its implementation which is composed of three parts: a linear feedback shift register (LFSR), a non-linear shift register (NFSR), and an accumulator to keep the output.

The present state of the LFSR is a linear function of its previous state. The generated LFSR sequence will be fed into the NFSR. In the NFSR, the present state is a non-linear function of its previous state. This non-linear function is composed of a linear function and a bent function. The LFSR and NFSR jointly build up a pseudo random number generator (PRNG). The output of this PRNG, $s(x)$, is the result of performing bitwise addition of the LFSR with the result of the NFSR function which feeds back into the NFSR as an input. This output is dependent on the initialized value of these two registers. Therefore, any change at the initialization value will cause the generated output sequence, $s(x)$, to be different. The value of the NFSR updates the accumulator. The accumulator is a register whose bits are XOR-ed with

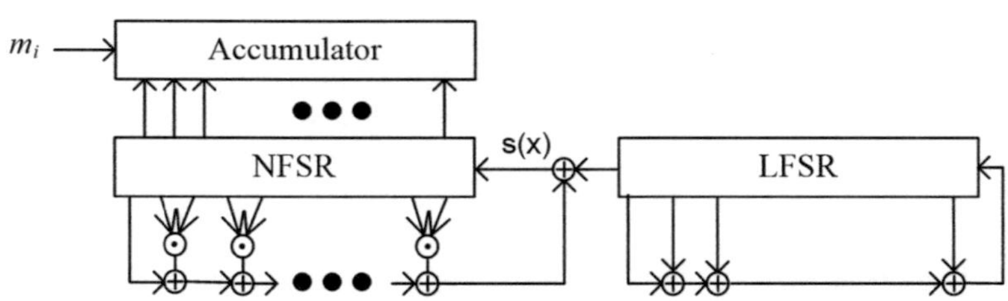

Fig. 4.7 The hardware for MAC generation as proposed in [1]

the value of the NFSR if $m_i = 1$. The input m_i is the input message bit which is checked by the accumulator on a bit-by-bit basis.

At the beginning of the process of MAC generation, the LFSR and NFSR are initialized with the authentication key while the accumulator is set to be zero. After initializing the registers, the message will be entered bit-by-bit at each clock cycle. If the input bit is one, then the accumulator will be XOR-ed with the content of the NFSR. Otherwise, nothing will happen. Both the LFSR and NFSR registers will be updated at each clock cycle. This process repeats until all bits of the message are checked by the accumulator. Therefore, the time required to generate the MAC is dependent on the length of the message and takes m clock cycles where m is the length of the message. The pseudo code of such an algorithm is presented in Algorithm 3.

The MAC algorithm in Algorithm 3 has a weakness when the message is a zero string which generates zero MAC as well. To overcome this flaw, a one-bit pad with value of one is deliberately appended to the end of the message and then this message is applied to the MAC algorithm.

Preventing collision at the MAC is very important in security strength of RBS cipher. If two or more different plaintexts have the same MAC as redundant bits then they will have the same keystream. Thus, the attacker can find some locations of the altered plaintext by performing bitwise addition of the ciphertexts of their plaintexts. However, the generated keystream is not required to be collision-free. Since two plaintexts are not the same, by performing the bitwise addition of them with plaintexts, two different altered plaintexts will result.

Hash collision probability for universal hash function is proved to be equal or less than the bias, ϵ, if the key is refreshed after each communication [7]. If L is the length of the input plaintext and w is the length of the NFSR, then the bias ϵ is defined for the MAC algorithm in [1] as:

$$\epsilon = \frac{L}{2^w} \tag{4.6}$$

Algorithm 3 Authentication algorithm

Initialize:

NFSR, LFSR = authentication key

Accumulator = 0

Process:

for every m_i in the message m **do**

    if $m_i = 1$ **then**

        Accumulator = Accumulator $\oplus$ NFSR

    end if

end for

Result:

Accumulator contains the MAC code

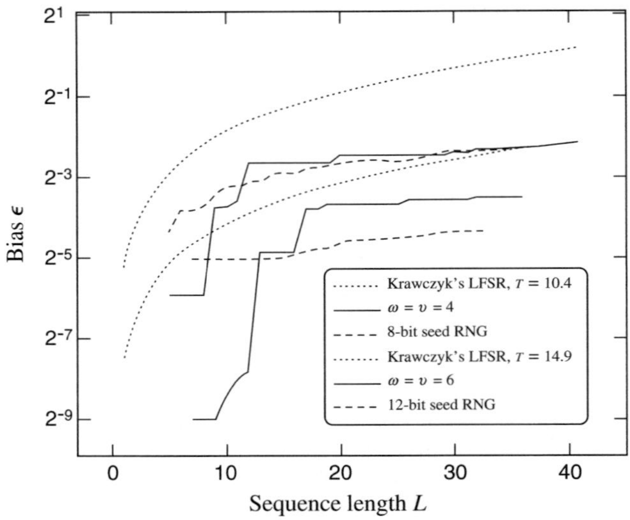

Fig. 4.8 The bias as it develops for growing sequence lengths obtained from the data in [1]

However, the experimental results in Fig. 4.8 show that the calculated bias is lower than the calculated results given by Eq. (4.6) as this equation gives only an upper bound of the bias. In Fig. 4.8, two different tag sizes w have been studied in [1] and their respective biases are plotted with solid lines. The dotted lines give the biases for the LFSR construction using equal amounts of randomness. The dashed lines show the behavior of random number generators. Lower curves give lower biases which offer lower probability of experiencing collision. As this figure shows, the obtained biases are lower than the expectation calculated by Eq. (4.6). For example, when w =6 and $L = 32$, the obtained bias is less than 2^{-3} which is far less than the calculated bias, $\epsilon = 32/2^6 = 2^{-1}$.

Based on Eq. (4.6), the probability of collision for RBS-132 is 2^{-62}. However, based on the experimental results for shorter plaintexts, it is expected that this probability will be lower than 2^{-62} which guarantees that encountering collisions in the redundant data will be very low and very close to zero.

4.5.3 Adapting the Chosen MAC to RBS

The first step for adapting the aforementioned authentication algorithm with RBS cipher is defining the size of the NFSR, the LFSR and the accumulator registers as well as the authentication key. The size of the accumulator and the NFSR are equal to the length of the MAC. Since the length of redundant data in RBS is m bits, two m-bit registers are reserved as NFSR and accumulator. Equations (4.7)–(4.10) presents the proposed NFSR functions for different designs of RBS denoted as $f(x)$.

RBS-83:

$$f(x) = 1 + x^{11} + x^{24} + x^{34} + x^{43} + x^{14}x^{19} + x^{20}x^{42} + x^{26}x^{29} + x^{37}x^{38} \quad (4.7)$$

RBS-100:

$$f(x) = 1 + x^{13} + x^{15} + x^{29} + x^{41} + x^{52} + x^{17}x^{24} + x^{25}x^{50} + x^{26}x^{27} + x^{32}x^{35} + x^{44}x^{45} \quad (4.8)$$

RBS-116:

$$f(x) = 1 + x^{15} + x^{17} + x^{33} + x^{47} + x^{60} + x^{20}x^{27} + x^{28}x^{58} + x^{29}x^{31} + x^{37}x^{41} + x^{52}x^{53} \quad (4.9)$$

RBS-132:

$$f(x) = 1 + x^{8} + x^{25} + x^{38} + x^{64} + x^{68} + x^{5}x^{14} + x^{20}x^{30} + x^{34}x^{41} + x^{46}x^{54} + x^{51}x^{60} \quad (4.10)$$

The adopted MAC algorithm is originally designed for stream ciphers [1]. The LFSR plays a major role in its authentication process because its present state will be referred for refreshing the authentication key in the next communication step. Since RBS is block cipher algorithm, and it uses fixed authentication key for each communication step, keeping the LFSR register is not required anymore. However, the LFSR key is required in generating pseudo-random numbers. Therefore, the LFSR key enters to NFSR register bit-by-bit.

In order to have the same key for both authentication and encryption, the size of the LFSR key is defined to be n bits which combined with the m-bit NFSR key to form a $n + m$-bit key [Eq. (4.11)]. Before applying the authentication key to the MAC generator, the key will be initialized once when the cipher starts up or the key changes. For initialization, the key is loaded into the NFSR while the input message remains zero during the process. The result after $2m$ clock cycles will be ready in the NFSR and it will be kept as a NFSR key.

$$\begin{aligned} K_{authentication} &= \{K_{LFSR}, K_{NFSR}\} \\ K_{LFSR} &= \{K_0, \ldots, K_{n-1}\} \\ K_{LFSR} &= \{K_n, \ldots, K_{n+m-1}\} \end{aligned} \quad (4.11)$$

Universal hash functions are guaranteed to be collision free if the key is refreshed after each usage [4]. To refresh the authentication key, the authentication key must be unique for each message. In other words, there must be unique K_{NFSR} and K_{LFSR} for each message. In order to refresh these keys in the RBS algorithm, they are defined as a function of the initial key and the plaintext data. One straightforward solution is through performing bitwise addition of the plaintext and the initial key value for generating authentication key per each plaintext as illustrated in Fig. 4.9.

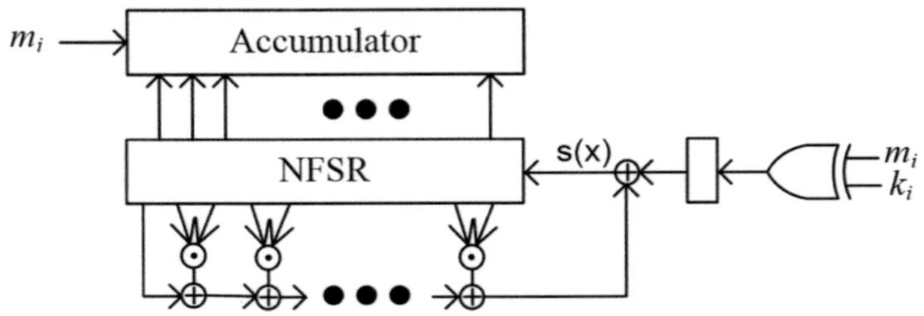

Fig. 4.9 Adapted MAC generator for RBS

The LFSR key is generated by performing bitwise addition of the plaintext bits m_i and the LFSR key k_i.

At this phase of generating the keystream, the size of the LFSR key is less than the size of the input. Therefore, the LFSR key is repeated from the beginning until it reaches the end in order to support generating the sequence of $s(x)$.

To prevent having zero as the MAC, the initialized value of both the NFSR and LFSR registers are required to be non-zero. To guarantee that their initialized value will not be zero for any message, the register placed after the XOR of LFSR key and the message is initialized with one. In the first clock cycle, if all bits of the NFSR are zeros, one bit with value "1" will be generated as the first bit of $s(x)$ and entered to the NFSR. Therefore, there is at least one bit with value "1" at the NFSR register which prevents generating zero as a MAC.

Generating the redundant bits takes $n + 1$ clock cycles since the size of the input of MAC is n-bit plaintext plus one bit padding. Generating the keystream requires m clock cycles because the length of the redundant bits input to the MAC is m bits without any padding. To generate the keystream, padding is not necessary since the value of redundant bits is always nonzero.

4.5.4 Encryption

The encryption process completes in two phases. In the first phase, the plaintext is altered through bitwise addition with the keystream. For the sake of area efficiency, the MAC generator circuit (Fig. 4.9) is used for altering the plaintext as well. This way, the NFSR and accumulator are loaded with the keystream and the message, respectively, while the input m_i is set to be one. The altered plaintext will be generated and stored in the accumulator in just one clock cycle.

In the second phase, the altered message is merged with the redundant data based on the secret key during data transmission. Figure 4.10 illustrates the process, where the altered plaintext bit (p_i), redundant bit (r_j), and encryption key bit (k_l) enter the cipher on a bit-by-bit basis. Depending on the value of key (k_l) either p_i or r_j will be transmitted.

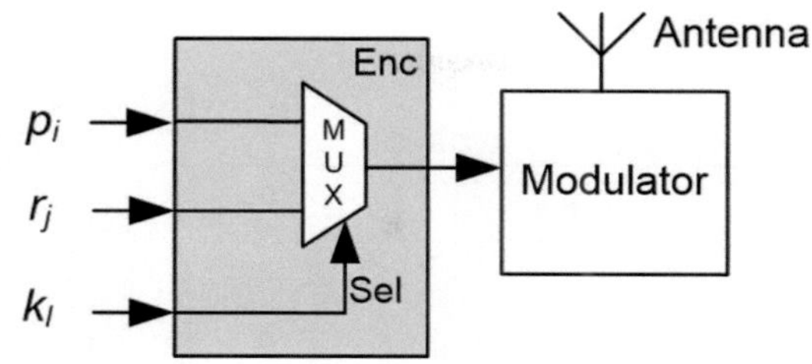

Fig. 4.10 The encryption module in the transmission process

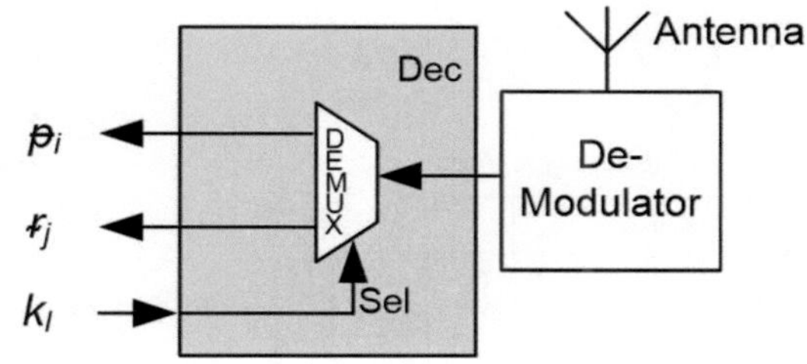

Fig. 4.11 Extracting the altered plaintext and the redundant data from ciphertext

4.5.5 Decryption

The decryption process completes in three phases. In the first phase, the redundant bits and the altered plaintext will be extracted from the received ciphertext (Fig. 4.11). Receiving data from the antenna and demodulating it, the received bit will be considered as either the altered plaintext bit, $\underline{p_i}$, or the redundant bit, $\underline{r_j}$, based on the value of the key k_l. These bits are shifted to their corresponding registers as they are received.

In the second phase, the keystream will be regenerated using the extracted redundant bits and the key. Performing bitwise addition of the regenerated keystream and the altered plaintext data, the original plaintext will be recovered as illustrated in Fig. 4.6b.

In the last phase, the redundant data is regenerated by calculating the MAC of the recovered plaintext as depicted in Fig. 4.6b. Comparing the received redundant data with the regenerated redundant data will authenticate the received message. In the case of failure in the authentication process, the decryption part returns a string of zeros as the decrypted message. This way, the algorithm would be secure against chosen ciphertext attacks which will be discussed in the next chapter.

4.5.6 Reception/Transmission

Figure 4.12 displays the encryption and decryption processes jointly with the data reception and transmission. Since the system is half-duplex, the reception and transmission will not happen at the same time.

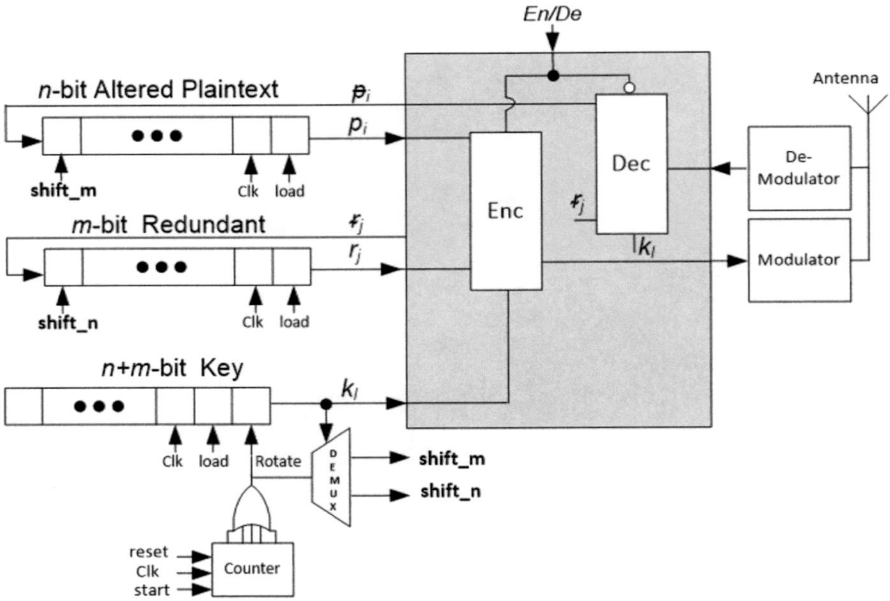

Fig. 4.12 Cipher plus transmitter and receiver

The En/De signal determines which process is being performed now, either encryption or decryption. The reception/transmission part is composed of a counter, a multiplexer and two registers. The registers store the sent or received message during transmission or reception. Here, there are two registers, one for the altered plaintext and the other for the redundant data. The counter keeps the number of bits required to be shifted to the encryption module or from the decryption module which is $n + m$ bits at the beginning of the each process. Since the total number of shifted bits is always fixed and shifting them is controlled by the secret key, there is no need to have separate counters to keep the number of shifted plaintext and redundant bits. The multiplexer is responsible for selecting which register is being shifted now based on the secret key. This multiplexer is designed such that it is active as long as the counter is working. Sending and receiving messages serially is an essential part of each RFID tag and is not designated for only this algorithm. Therefore this part is not counted in calculating the experimental results such as the area and the power consumption except the ciphers and multiplexer.

To send out the encrypted message, the counter will first be initialized with the total number of bits required to send which is $n + m$ in the RBS algorithm. Then based on the key, the least significant part of the corresponding register will be shifted to the encryption module. The corresponding bit from this module will be sent to the modulator to be transmitted. This process continues until the counter becomes zero. At this time, all of the altered plaintext and redundant bits are sent over the air.

To receive the encrypted message, the same process will be followed with the only difference that the received bits will be shifted to its own corresponding register based on the secret key. At the end of the process, all the received bits are separated by the decryption module and stored in their registers.

The transmission and reception algorithms are illustrated by the pseudo code in Algorithms 4 and 5, respectively, where m is the plaintext data length and n is the redundant data length. Both algorithms are composed of trivial shifting and selection operations which allows the system to encrypt or decrypt the data during sending and receiving data. This capability makes the RBS algorithm very efficient in terms of timing overhead. The only considerable overhead in the encryption and decryption processes is the MAC implementation which will be discussed in detail in the experimental results chapter.

Algorithm 4 RBS transmission algorithm

```
counter = 0
for i in range (n + m) do
    if key[i] = 0 then
        send data to MSB(plaintext)
        shift right plaintext register
    else
        send data to MSB(redundant)
        shift right redundant register
    end if
    shift right key register
end for
```

Algorithm 5 RBS reception algorithm

```
counter = 0
for i in range (n + m) do
    if key[i] = 0 then
        shift right plaintext register
        send LSB(plaintext) to Enc module
    else
        shift right redundant register
        send LSB(redundant) to Enc module
    end if
    shift right key register
end for
```

4.6 Overall RBS System

The RBS algorithm performs the encryption and decryption processes along with authentication using the same hardware. Since RFID systems are half-duplex, either encryption or decryption will be performed at a given instant, while authentication is a part of both processes—not an optional service.

Figure 4.13 describes the overall system with a flowchart. In the encryption mode, the MAC of plaintext and redundant bits is generated first. In this part, first the MAC generator is initialized with the XOR of the plaintext and the secret key. Then, the plaintext is entered at each clock into the MAC generator bit-by-bit. It repeats for $n + 1$ clock cycles where n is the size of plaintext. The next step is generating the keystream which is similar to the last step except that it repeats for m cycles where m is the number of redundant bits. Then the result of the bitwise addition of the keystream and the plaintext is calculated as the altered plaintext. The last step is transmitting the message. In this step, either the altered plaintext or the redundant bits will be sent out based on the secret key.

In the decryption mode, the altered plaintext is first separated from the redundant bits based on the key secret at the receiver side and shifted into their own registers as the receiver is receiving the message bit-by-bit from the demodulator. Given knowing the redundant bits, the keystream is generated by calculating the MAC of the redundant bits similar to the encryption mode. Afterward, the plaintext is found by executing bitwise addition of the keystream and the altered plaintext. Eventually, the redundant bits are regenerated—using the plaintext—and compared with the received redundant bits. If these two are identical, then the message will be authenticated. Otherwise, the message will be known as a corrupted message and discarded.

4.7 Conclusion

This chapter was dedicated to describing the RBS algorithm which is a new authenticated symmetric encryption method for RFID systems based on inserting redundant bits into the original data bits. The proposed method provides authentication, integrity, and confidentiality, all together. The security level of the proposed system can be adjusted without changing the underlying MAC algorithm just by changing the number of redundant bits and the plaintext.

The RBS algorithm is a lightweight cryptosystem with performance, cost overhead, and security strength that makes it a good fit for adoption by RIFD system used in the Internet of Things (IoT). Compared to other cryptosystems, the only disadvantage of the RBS algorithm is the length of the ciphertext which is longer than others. However, if existing cryptosystems are to provide authentication, the length of RBS ciphertext will be comparable with the length of their ciphertexts especially for stream ciphers which the authentication part is recommended.

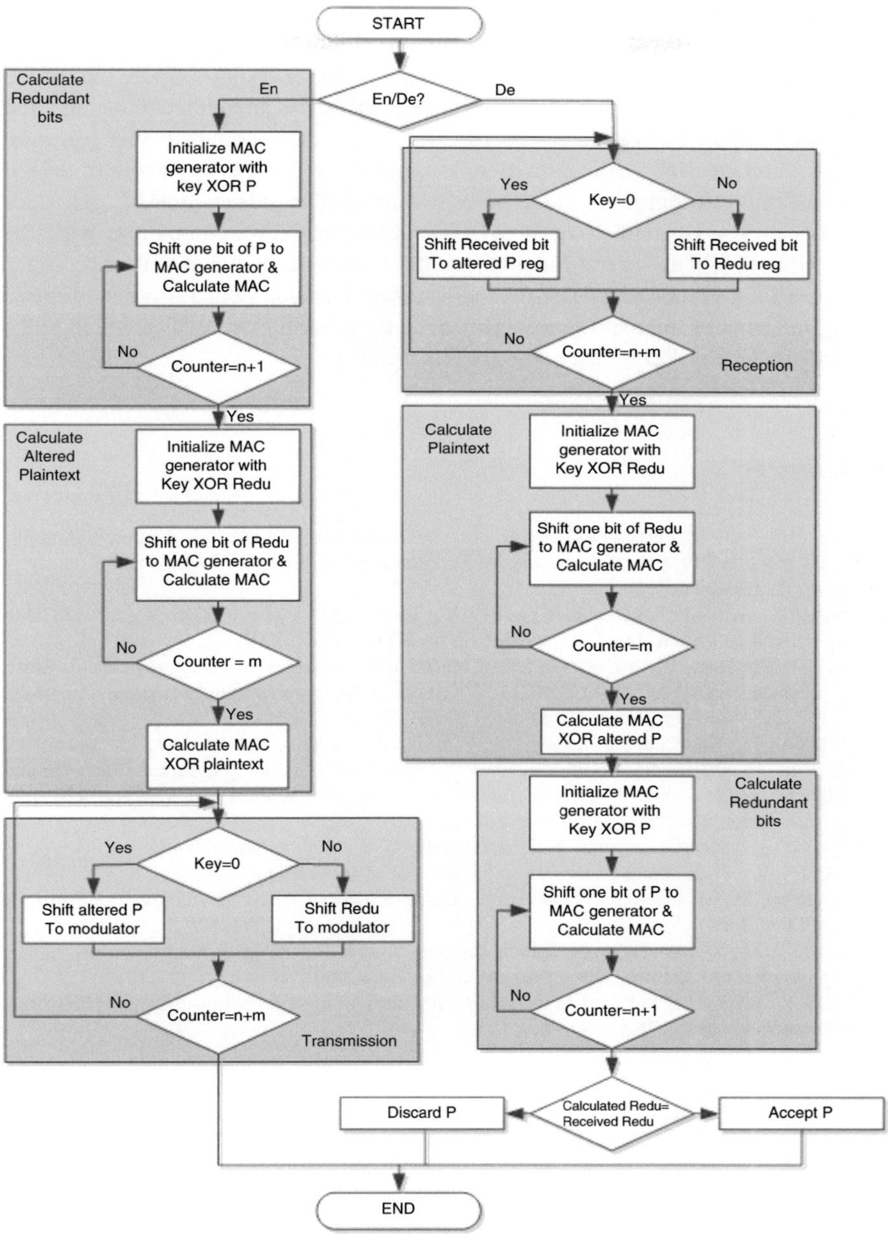

Fig. 4.13 The flowchart of the RBS algorithm of the overall system

At the end of this chapter, the hardware implementation of the RBS algorithm is explained. The RBS implementation is composed three parts: the MAC generator that produces the redundant bits and the keystream, the encryption cipher embedded in the sender to merge the altered plaintext with the redundant bits, and the decryption cipher embedded in the receiver to separate redundant bits from the altered plaintext. The main part of this hardware is the MAC generator which consumes more resources than the other two parts. The presented MAC generator for RBS cipher is adapted from [1]. To make this MAC compatible with the proposed RBS cipher, several modification has been done on the initialization phase and the LFSR component. These modifications made the considerable resource saving in terms of area and power. However, it imposed extra cycles which make the performance of RBS cipher slightly disgraced.

References

1. Agren, M., Hell, M., Johansson, T.: On hardware-oriented message authentication with applications towards RFID. In: Proceedings of International Workshop on Lightweight Security & Privacy (LightSec) (2011)
2. Agren, M., Hell, M., Johansson, T., Meier, W.: Grain-128a: a new version of grain-128 with optional authentication. Int. J. Wire. Mob. Comput. **5**(1), 48–59 (2011)
3. Black, J., Rogaway, P.: CBC MACs for arbitrary-length messages: the three-key constructions. In: Advances in Cryptology - CRYPTO 2000. Lecture Notes in Computer Science, vol. 1880, pp. 197–215. Springer, Berlin, Heidelberg (2000)
4. Bogdanov, A., Knudsen, L., Leander, G., Paar, C., Poschmann, A., Robshaw, M., Seurin, Y., Vikkelsoe, C.: PRESENT: an ultra-lightweight block cipher. In: Cryptographic Hardware and Embedded Systems - CHES 2007. Lecture Notes in Computer Science, vol. 4727, pp. 450–466. Springer, Berlin, Heidelberg (2007)
5. Engels, D., Saarinen, M., Smith, E.: The Hummingbird-2 lightweight authenticated encryption algorithm. In: Proceedings of Workshop on RFID Security (RFIDSec) (2011)
6. Hankerson, D., Menezes, A., Vanstone, S.: Guide to Elliptic Curve Cryptography. Springer, Heidelberg (2004)
7. Nguyen, L.H., Roscoe, A.W.: New combinatorial bounds for universal hash functions. IACR Cryptology ePrint Archive (2009). http://eprint.iacr.org/2009/153
8. Yuksel, K., Kaps, J., Sunar, B.: Universal hash functions for emerging ultra-low-power networks. In: Proceedings of CNDS (2004)

Chapter 5
RBS Security Analysis

Abstract One of the most important factors of a highly secure cryptosystem is its resilience against security attacks. There exists several well known attacks that target RFID systems. Having a lightweight cipher in terms of area and power consumption must not compromise its resilience to such security attacks. Otherwise, the advantages of such a lightweight cipher in terms of hardware implementation will not be valued. In Chap. 4, the RBS algorithm and its hardware implementation were introduced and discussed in details. In this chapter, the security of the RBS algorithm is investigated against several powerful and well-known attacks such as the known-plaintext attack, chosen-plaintext attack, chosen-ciphertext attack, differential attack, substitution attack, related key attack, linear cryptanalysis algebraic attack, cube attack and side channel attack. We show how the RBS algorithm is resilient against these attacks.

One of the most important factors of a highly secure cryptosystem is its resilience against security attacks. There exists several well known attacks that target RFID systems. Having a lightweight cipher in terms of area and power consumption must not compromise its resilience to such security attacks. Otherwise, the advantages of such a lightweight cipher in terms of hardware implementation will not be valued.

In Chap. 4, the RBS algorithm and its hardware implementation were introduced and discussed in details. In this chapter, the security of the RBS algorithm is investigated against several powerful and well-known attacks such as the known-plaintext attack, chosen-plaintext attack, chosen-ciphertext attack, differential attack, substitution attack, related key attack, linear cryptanalysis algebraic attack, cube attack and side channel attack. In what follows, we show how the RBS algorithm is resilient against these attacks.

5.1 Security Model

A typical RFID system consists of one eligible reader and N RFID tags. There is a unique key for each tag that is shared with the authorized reader. Consider the scenario where a sample tag A sends its encrypted message through ciphertext C to

A. Khattab et al., *RFID Security*, Analog Circuits and Signal Processing,
DOI 10.1007/978-3-319-47545-5_5

an eligible reader B. Listening to all the communications on the channel and/or modifying them, an adversary ε tries to discover the encryption key using the following operations:

- Eavesdropping: The adversary ε captures the signals transmitted between A and B, then demodulates and decodes the received signals to extract C.
- Modification: The adversary ε modifies the signals on the channel in order to (1) alter the data through bit flipping (converts a "0" bit into "1" or vice versa), (2) append bits to the original data bits, and (3) delete some of the data bits.

Changing the length of the data by ε in this security model, either by appending some bits or deleting some of the bits, invalidates the ciphertext C in the RBS system as the size of the ciphertext is fixed. Therefore, this issue will not be discussed here.

5.2 Mathematical Background

In the RBS algorithm, the plaintext $P \in F_2^n$ and the encryption key $K \in F_2^{n+m}$ are used together to generate the ciphertext $C \in F_2^{n+m}$ where n is the length of plaintext and m is the length of redundant data.

The Key K is divided into two sub-keys $K_{NFSR} \in F_2^m$ and $K_{LFSR} \in F_2^n$.

$$\begin{aligned} K_{authentication} &= \{K_{LFSR}, K_{NFSR}\} \\ K_{LFSR} &= \{K_0, \ldots, K_{n-1}\} \\ K_{LFSR} &= \{K_n, \ldots, K_{n+m-1}\} \end{aligned} \tag{5.1}$$

Let $r_d \in F_2^m$ represent redundant data in the ciphertext. Then

$$r_d = \mathrm{MAC}_{K_A}(P) \tag{5.2}$$

where

$$K_A = \{(K_{NFSR} \oplus P), g(K_{LFSR}, P)\} \tag{5.3}$$

Let $k_s \in F_2^n$ denote the keystream. Then

$$k_s = \mathrm{MAC}_{K_B}(r_d) \tag{5.4}$$

where

$$K_B = \{(K_{NFSR} \oplus r_d), g(K_{LFSR}, r_d)\} \tag{5.5}$$

The altered plaintext, $a_p \in F_2^n$, is generated by bitwise addition of the keystream k_s and the plaintext P.

$$a_p = k_s \oplus P \tag{5.6}$$

The ciphertext is generated by merging r_d with a_p using the encryption key K, that is, $C = E_K(r_d, a_p)$. In this security model, tag A encrypts its message P with the encryption key K and sends it to the reader B using above equations.

$$A \rightarrow B : \{\mathrm{MAC}_{K_A}(P), \mathrm{MAC}_{K_B}(\mathrm{MAC}_{K_A}(P)) \oplus P\}_K \tag{5.7}$$

Assuming that the MAC encrypts the message with K, the following Lemmas hold true if ε does not have K.

- **Lemma 1-** If the attacker ε does not have the key K, it cannot retrieve the corresponding redundant data r_d for the plaintext P as $r_d = \mathrm{MAC}_{KA}(P)$.
  Proof: To find the MAC of any message, having both the key and the message is required. In RBS terminology, $r_d = \mathrm{MAC}_{KA}(P)$ *and since the attacker does not have the key, it cannot find the redundant data of that plaintext.*
- **Lemma 2-** If the attacker ε does not have the key K, it cannot retrieve the corresponding keystream k_s for the plaintext P.
  Proof: Based on Lemma 1, ε *cannot find* r_d *from P so it cannot find* $k_s = \mathrm{MAC}_{KB}(r_d)$
- **Lemma 3-** If the attacker ε does not have the key K, it cannot retrieve the altered plaintext a_p from the plaintext P.
  Proof: Based on Lemma 2, k_s *is not revealed through P, and hence,* $a_p = k_s \oplus P$ *is not revealed as well.*
- **Lemma 4-** If the attacker ε does not have the key K, he cannot retrieve keystream k_s from redundant data r_d as $k_s = \mathrm{MAC}_{KB}(r_d)$.
  Proof: It is proved similar to Lemma 1.
- **Lemma 5-** If the attacker ε does not have the key K, it cannot retrieve the altered plaintext a_p from the redundant data r_d.
  Proof: Based on Lemma 2, ε *cannot find* k_s *from* r_d*, and consequently, it cannot find* a_p *since in* $k_s \oplus P$, k_s *is unknown for* ε.

The MAC is a one-way encryption algorithm and having the digest/key combination, it is practically impossible to retrieve the plaintext. The following Lemmas hold true over this fact if ε does not have access to the key K:

- **Lemma 6-** plaintext P cannot be retrieved from redundant data r_d.
- **Lemma 7-** plaintext P cannot be retrieved from keystream k_s.
- **Lemma 8-** plaintext P cannot be retrieved from altered plaintext a_p.
- **Lemma 9-** redundant data r_d cannot be retrieved from keystream k_s.
- **Lemma 10-** redundant data r_d cannot be retrieved from altered plaintext a_p.
- **Lemma 11-** if ε does not have the K, he cannot retrieve neither r_d nor a_p from Ciphertext C.

Proof: Ciphertext C is a mixture of redundant data r_d, and altered plaintext a_p. Based on Lemmas 5 and 10, r_d is not revealed from a_p and vice versa. Therefore, ε may try arbitrary combinations of r_d and a_p in the ciphertext C. Although there is dependency between r_d and a_p, ε cannot retrieve a_p, even if it has r_d, given Lemma 3.

5.3 RBS Security Against Common Attacks

In the rest of this chapter, the security of the RBS algorithm is evaluated against powerful and well-known attacks in cryptanalysis including, but not limited to, the known plaintext attack, the chosen-plaintext attack and the chosen-ciphertext attack.

5.3.1 *Brute Force Attack*

The security strength of a cryptographic algorithm depends on how difficult it is to break an encrypted message without the knowledge of the encryption key used by the algorithm. Unconditional security would be perfect, however, the only known such cipher is the one-time pad. For all other encryption algorithms, providing the assumed computational security is possible if it either takes too long or is too expensive to break the cipher, assuming that the attacker has access to reasonable computing resources and time to break the cipher. The cryptographic strength may also depend on other parameters such as the worth of the message, the lifetime of the message privacy, etc. The best parameter for measuring the security strength can be the number of possible keys which the adversary has to try to retrieve the right key.

The brute-force attack is a powerful attack which involves trying all of the possible keys until an understandable relation between the ciphertext and its plaintext is obtained based on that key. The length of key used in the cipher determines the security level of the cipher against the brute-force attack. Increasing the length of keys will increase the time required for breaking the cipher exponentially. Therefore, a cipher with n-bit key can be broken in a worst-case time proportional to 2^n. However, half of all the possible keys must be tried on average in order to accomplish success because of weak keys.

Table 5.1 shows how much time is required to perform a brute-force attack for various common key sizes when either a single decryption process or a million parallel processes are used. In Table 5.1, keys with a size of 128 bits or longer provide enough security for contemporary cryptosystems. However, for RFID systems the length of the key used in the lightweight encryption algorithms is between 80 bits (e.g. PRESENT) to 128 bits because of the trade-off between security, cost and performance. To have a higher security with keys longer than 128 bits, the cost in terms of area and power will become higher while the performance

Table 5.1 Time required for breaking key by the brute-force attack

Key size [bits]	Key space	Time required at 1 decryption/μs	Time required at 10^6decryption/μs
32	2^{32}	35.8 min	2.15 ms
56	2^{56}	1142 years	10.01 h
80	2^{80}	1.9×10^{10} years	1.9×10^{4} years
128	2^{128}	5.4×10^{24} years	5.4×10^{18} years
168	2^{168}	5.9×10^{36} years	5.9×10^{30} years

will decrease. In the RBS algorithm, the length of the key can be varied between 83 and 132 bits which provides the same key space provided by keys with length of 80–128 bits.

5.3.2 *Known-Plaintext Attack*

In the known-plaintext attack, the attacker ε has a pair of valid plaintext/cipher (P/C) and tries to discover the key K. It can happen through eavesdropping on the channel between the tag and the reader when the tag is sending a special message that is likely ε has access to it such as the EPC number of the tag.

Based on Lemmas 1 and 3, ε cannot retrieve r_d and a_p from P, and hence, it cannot locate them inside C as well. To regenerate P through C, ε tries arbitrary combinations of $\dot{K}$, $\dot{r}_d$ and $\dot{a}_p$ from C, where $C = Merge_{\dot{K}}\{\dot{r}_d, \dot{a}_p\}$. Based on Lemma 6, P cannot be revealed from only r_d. Likewise, P cannot be revealed from only a_p based on Lemma 8. The only way to regenerate P is through the right $\{\dot{r}_d, \dot{a}_p, \dot{K}\}$ combination which satisfies $C = E_{\dot{K}}(\dot{r}_d, \dot{a}_p)$. Meanwhile, there are as many different combinations for $\dot{K}$ as the key space size in RBS cipher whereas only one $\dot{K}$ satisfies $C = E_{\dot{K}}(\dot{r}_d, \dot{a}_p)$. In other words, knowing one pair of P/C does not shrink the key space.

5.3.3 *Chosen-Plaintext Attack*

A chosen-plaintext attack is an attack model for cryptanalysis which assumes that the attacker has the ability of choosing arbitrary plaintexts to be encrypted and gain their corresponding ciphertexts [2]. The goal of the attacker is to find out further information which reduces the security strength of the encryption algorithm by comparing the ciphertexts and their plaintexts. In the worst case, the attacker may find the secret key by this attack.

Based on the definition of this attack, it is assumed that the attacker ε has access to the decryption device, and the attacker can encrypt its own plaintexts into their ciphertexts. According to the RBS algorithm, the attacker will have to find the location of the r_d bits or the a_p bits inside C by this information.

Suppose that ε has two pairs of (P_1/C_1) and (P_2/C_2) while $C_1 = Merge_k\{r_{d1}, a_{p1}\}$ and $C_2 = Merge_k\{r_{d2}, a_{p2}\}$. The attacker performs peer-to-peer comparison of the bits in C_1/C_2 for the purpose of extracting further information in order to reduce the key space.

Changing each bit of the plaintext reflects itself over some of the bits of the redundant data and the altered plaintext, but the number of changed bits as well as their location depends on the used MAC algorithm key which is also unique per each plaintext and is unknown to ε. Therefore, ε cannot predict any changes in ciphertexts of two plaintexts.

Comparing each pair of bits $C_1[i]$ and $C_2[i]$, they might be either equal or not. If $C_1[i] \neq C_2[i]$, this change might represent either a change in r_{d1} or a change in a_{p1}. Besides, for $C_1[i] = C_2[i]$ the ith bit might belong to r_d or a_p. For these reasons, no useful information about the key can be obtained by tracing the changes in C_1 and C_2.

Considering the plaintext data as well as the ciphertext data, changing just one bit of the plaintext changes about half of the ciphertext bits since r_d and a_p are generated through MAC and these changed bits are randomly distributed in the ciphertext based on the fact that the underling MAC is a PRNG. Besides, these changed bits may belong either to r_d or a_p. Since this is a very special case of differential attacks, we leave it here and discuss it in more detail later.

5.3.4 Chosen-Ciphertext Attack

A chosen-ciphertext attack is an attack model in which the attacker ε has the capability of decrypting its own ciphertexts and retrieve their corresponding plaintexts. Suppose that ε has captured a valid ciphertexts C through eavesdropping on the channel and has also decrypted its corresponding plaintext P. Since ε has access to decryption device, ε can modify some bits of $C = Merge_k\{r_d, a_p\}$ to interpret their reflection on its decrypted data. It must be noted that ε does not know K, a_p, and r_d based on Lemma 11.

For a particular modified bit, if it belongs to r_d, the decryption part will not authenticate C as r_d is unique per a_p. Likewise, if the modified bit belongs to a_p, C will not be authenticated as a_p is also unique per r_d. In both cases, the decryption part will return string of 0's which does not reveal any information about K.

In order to get a valid plaintext P from a modified C, the changed bits must belong to both r_d and a_p such that they must satisfy $a_p = \text{MAC}_{KB}(r_d) \oplus P$. On the other hand, the number of possible ciphertexts in RBS cipher is 2^{m+n} out of which 2^n can be authenticated and accepted at the receiver side where m and n are the length of the redundant data and the plaintext, respectively. This means that the probability of finding the right match of a_p and r_d is as low as 2^{-m}. Suppose even such a rare match occurs and ε collects valid ciphertexts $\{C_1, C_2\}$ and their corresponding plaintexts

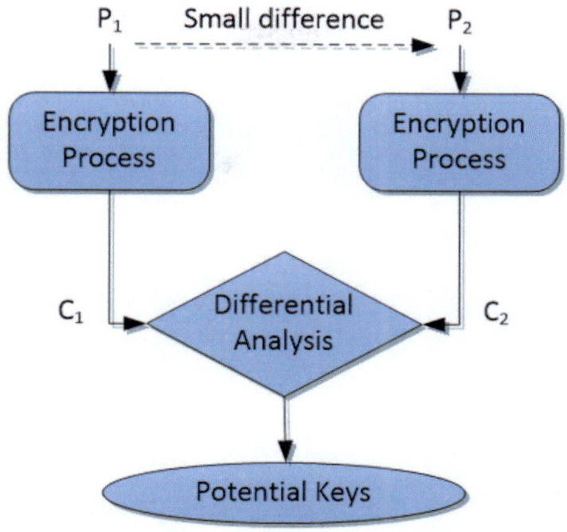

Fig. 5.1 Differential attack model

$\{P_1, P_2\}$. This scenario would be similar to chosen-plaintext attack in which finding that the modified bits belong to r_d or a_p is not practically possible. Consequently, having these data does not shrink the RBS key space.

5.3.5 *Differential Attack*

The differential attack is a chosen-plaintext attack which extracts the relationship between the difference of two inputs and the difference of their corresponding outputs. In this attack, the attacker searches for plaintext–ciphertext pairs whose difference is constant or non-random, and investigates the differential behavior of the cryptosystem with the hope that it may detect statistical patterns in the distribution of differences between the ciphertexts (Fig. 5.1). The difference between two plaintexts is very small and usually the difference is in just one bit.

If (P_1, C_1) and (P_2, C_2) are two pairs of different plaintexts and ciphertexts then $(\Delta P, \Delta C)$ states the difference between the two given pairs of plaintexts and ciphertexts where $\Delta P = P_1 \oplus P_2$ and $\Delta C = C_1 \oplus C_2$.

In order to be resilient against differential attacks, the cryptosystem should be a good pseudorandom number generator (PRNG). The underlying MAC implemented in RBS is based on the pseudorandom generator discussed in Chap. 4. This performance of random behavior of the used PRNG has been verified by simulating the output of the RBS cipher for a set of 64-bit input plaintexts where each two adjacent inputs in this set are different in only one bit. The RBS algorithm has been applied for 100 different keys per each input in the set. The results for the first ten keys are summarized in Table 5.2. Considering RBS-132, the 64 generated 132-bit ciphertexts for 64 input plaintexts in the set as a two dimensional array of 132×64, we have investigated the effect of a single bit change in all of the 132 columns and 64 rows of this array.

Table 5.2 Simulation of RBS outputs when the inputs are different in one bit

	Redundant data (# bits: 68)				Altered plaintext (#bits: 64)				
	Column			Row	Column			Row	Total
Key	Max Transition	Min Transition	Average Transition	%Changed bits	Max Transition	Min Transition	Average Transition	%Changed bits	%Changed cipher
1	41	23	31.06	51%	42	23	31.53	50%	50%
2	41	21	30.57	51%	41	22	31.02	49%	49%
3	40	22	33.46	50%	41	23	32.2	50%	50%
4	40	25	32.16	49%	37	24	31.8	48%	49%
5	39	21	32.82	50%	44	22	32.47	51%	50%
6	45	23	32.94	52%	39	24	31.73	51%	51%
7	39	23	31.6	48%	41	24	32.52	50%	48%
8	43	22	31	48%	41	21	31.42	50%	49%
9	42	22	32.44	50%	41	20	31.58	50%	50%
10	40	22	31.81	51%	41	21	31.66	49%	49%

Regarding the changes in the columns of this array, Table 5.2 lists the minimum, average, and maximum transitions in the ciphertext bits by changing only one bit of the plaintext. For instance by applying Key 1, on average 31.06 transitions has occurred for each bit of the redundant data. In other words, a single bit change in the plaintext will change each bit of the redundant data with a probability of 31.0/64 $\approx$ 0.49. A very similar results were obtained for the altered plaintext.

Regarding the rows of the array, the average number of redundant bits and altered plaintext bits that are different from the output of the previous input is also presented in Table 5.2. The reflected changes in the redundant part and the altered plaintext part of the ciphertext are again very close in value to each other. As the last column of the table shows, just a single bit change in the plaintext transforms almost half of the ciphertext bits uniformly in both the redundant part and the altered plaintext part of the ciphertext.

Performing the simulation test for special case input plaintexts of all 0's or all 1's almost, the same results have been obtained. This simulation practically confirms the good pseudorandom behavior of RBS which is a prerequisite for being resilient against differential attacks.

5.3.6 Substitution Attack

The substitution attack is especially introduced for stream ciphers wherein the ciphertext is the result of performing bitwise addition of the keystream and the plaintext. In this attack, the attacker tries to replace a legitimate pair of message and MAC (m, t) with its own pair of message and MAC (m', t') and succeeds with probability P_S. Therefore, if the attacker can find such a pair, it can send its own message while it will be authenticated in the reception side.

It has already been proven in [1] that the utilized MAC generator in RBS is resilient against this attack and the probability of substituting the pair of message and MAC is very low as long as the message and MAC are separated and can be distinguished in the ciphertext. This probability in RBS will be even less since the attacker cannot even distinguish the message from its MAC in the output based on the Lemmas 1 and 3.

5.3.7 Related Key Attack

The related key attack was first introduced in [3]. In this attack, the attacker ε is assumed to be able to obtain the ciphertext from the cipher for different plaintexts under different keys K_1, K_2, etc. The values of these keys are initially unknown for the attacker. Using these data, the attacker tries to look for some information about the secret keys by observing the operation of the cipher under several different keys for the same plaintexts and finding the relationship between their output ciphertexts.

One possible scenario to test RBS against this attack could be by assuming that the attacker gives the same plaintext P to different tags which have different keys (K_1, K_2) and tries to analyze the generated ciphertexts (C_1, C_2) for that particular message P.

As discussed before, the redundant data r_d that is the MAC generator's output is computed as a function of the plaintext P and the key K. Consequently, $r_{d1} \neq r_{d2}$ as $K_1 \neq K_2$. Since the attacker does not have access to K_1 and K_2, it cannot find the corresponding r_{d1} and r_{d2} based on Lemma 1. The same condition applies for the altered plaintexts a_{p_1} and a_{p_2} based on Lemma 3. Therefore, the attacker cannot extract (r_{d1}, a_{p_1}) and (r_{d2}, a_{p_2}) from ciphertexts C_1 and C_2, respectively. In other words, the attacker cannot extract any useful information to shrink the key space size of RBS.

Similar simulations as those performed for differential attacks have been performed for the related key attack with the difference that for the same plaintext, ciphertexts are generated for secret keys that are different in one or more bits. The results of these simulations confirm that the outputs of the RBS cipher have random behavior by changing the key as we expected.

The keys used for encryption/decryption and authentication are the same in RBS ciphers. Therefore, any changes in the key will change both the value of the redundant bits and the altered plaintext along with changing their locations in the ciphertext. Therefore, different ciphertexts will be generated by applying different keys which makes the cipher resilient against related-key attacks. Having different keys for encryption and authentication may make the RBS cipher vulnerable to related-key attacks unless the dedicated authentication key becomes XOR-ed with the encryption key before applying it to the MAC generator. Thus, the applied key to the MAC will be a function of encryption key and the authentication key. Any changes in any of keys will affect the entire ciphertext.

Apart from comparing the output of the RBS cipher for different valid secret keys, the attacker may modify the secret key partially and encrypt the message under an invalid modified key while the key remains secret for the attacker. For example, there is a possibility in RBS that filliping any bits of the secret key may change the balance of the number of redundant bits and altered plaintext bits in the ciphertext. If it does not change the balance in the key, the message will be authenticated in the receiver side (if attacker has modified the key at both the sender and receiver sides). In this case, those flipped bits in the key are complement of each other. If modifying the key changes the balance of the encryption key, the message will not be authenticated in the receiver side since the receiver has not received all bits of either the altered plaintext or the redundant bits. Therefore, authenticating the message upon reception helps the attacker to find the relationship between the bits of the key.

Preventing this attack is easy through error correction of the secret key which can be performed by adding two counters. These counters are responsible for counting the number of redundant bits and altered plaintext bits inside the ciphertext (Fig. 5.2) and checking the balance of the ciphertext when the encrypted message is sent out

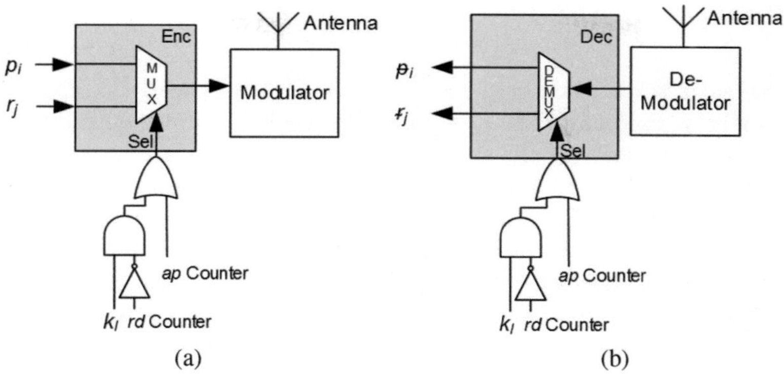

Fig. 5.2 Error correction of the secret key. (**a**) Transmitter side. (**b**) Receiver side

or being received. If one of the counters reaches its maximum pre-defined number, it will dominate the select pointer to correct selecting the inputs/outputs in the encryption/decryption process and fix the balance of the redundant bits or the altered plaintext at the rest of the ciphertext.

5.3.8 Linear Cryptanalysis

The linear attack is a known plaintext attack which was introduced for the first time in [11]. In this attack, the attacker tries to find linear expressions involving some bits of the plaintext, ciphertext and secret key. This expression can be stated as follows:

$$X_i + X_j + \ldots + Y_a + Y_b = K_m + \ldots + K_n \tag{5.8}$$

where X_i represents the ith bit of the plaintext, Y_a represents the ath bit of the ciphertext and K_m represents the mth bit of the secret key.

In RBS cipher, the key is applied after initialization for two rounds which passes through the nonlinear function of the NFSR. Also, during the process of generating the redundant data and the keystream, one pseudo random number is generated at each clock cycle and shifted into the NFSR. The value of each bit of the output is the result of bitwise addition of these random numbers AND-ed with the input bits. Since the coefficient of the random number for each bit of the output is different, factorizing the coefficients to find a linear function is not possible.

5.3.9 Algebraic Attack

An algebraic attack is a new cryptanalytic method which is designed especially for stream ciphers based on LFSRs. The main idea behind this method is discovering and solving a system of multivariate polynomial equations over a finite field. In this kind of attacks, the attacker has a plaintext and its corresponding ciphertext. Therefore, the attacker can find the corresponding keystream based on these two given inputs. Based on the obtained keystream, the attacker sets up a system of polynomial equations which entry is the keystream [5]. The goal is recovering the initial state of the LFSR which is the key. Let's assume one equation is found for the tth bit of the keystream as given by the following equation:

$$0 = f(k_t, \dots, k_{t+r-1}, z_t, \dots, z_{t+r-1}) \tag{5.9}$$

where z_i is the ith bit of the keystream and k_i is the ith bit of the secret key. Then, the same equation is correct for any clock:

$$\left\{\begin{array}{l} 0 = f(k_0, \dots, k_{r-1}, z_0, \dots, z_{r-1}) \\ 0 = f(k_1 + \dots + k_r, z_1, \dots, z_r) \\ 0 = f(k_2 + \dots + k_{r+1}, z_2, \dots, z_{r+1}) \end{array}\right\} \tag{5.10}$$

If the relationship between the states and the output bits can be stated as one multivariate equation of low degree without extra variables, then the cipher will be broken in polynomial time [5]. In block ciphers, attackers set up a matrix of multivariate functions which variables are the bits of the input plaintext, the ciphertext and the secret key. Solving this system of functions will lead to recovering the key.

The output of the RBS cipher is composed of a mixture of two strings: the redundant data, r_d and the altered plaintext, a_p. The redundant data is a function of the secret key and the input plaintext and it is independent of the altered plaintext which is a function of the keystream, k_s and the input plaintext while the keystream is a function of the redundant data. In this kind of attacks, the attacker is required to set up a system of equations for each of these two strings separately because each of these two strings has a different initial key and different inputs. Setting up these equations will not happen unless these strings have been already distinguished in the ciphertext. Meanwhile in known-plaintext attacks, it is proved that it is not possible for the attacker to find the redundant data and its corresponding altered plaintext in the output by having a pair of the plaintext and ciphertext.

5.3.10 Cube Attack

Cube attacks are chosen-plaintext attacks introduced for symmetric cryptosystems. The ciphertext bits produced by this algorithm are values of polynomials depending on initial vectors in stream ciphers and bits of a plaintext plus secret key for block ciphers [7]. In this attack, the attacker tries to obtain a linear equation of the secret key bits by combining the equations for an output bit of the cipher for a set of inputs and keys. A cipher is vulnerable to this attack if an output bit of the cipher can be represented as a sufficiently low degree polynomial over GF(2) of the key and input bits. This attack is a type of algebraic attacks which degree of variables in the equation is more than one. For example, in the following equation p is a polynomial of degree 3 with 5 variables:

$$p(x_1, x_2, x_3, x_4, x_5) = x_1x_2x_3 + x_1x_2x_4 + x_2x_4x_5 + x_1x_2 + x_2 + x_3x_5 + 1 \quad (5.11)$$

The success probability of the cube attack is high if the degree of the internal state transit function in a stream cipher is low. For example, Trivium is vulnerable to this attack because the degree of its internal state transit function grows slowly [6]. In RBS, resilience against this attack depends on the nonlinear function of the NFSR. This function can be introduced such that it provides a high degree of states.

5.3.11 Side Channel Attack

All of the investigated attacks so far were based on the analysis of attacker's knowledge about the security algorithms rather than the hardware to find their weak points. This information may consist of a set of plaintexts and their corresponding ciphertexts as the case in known-plaintext and chosen-plaintext attacks. Based on this information, the attacker tries to find a relationship between them and the secret key. In contrast, the attacker in a side channel attack relies on the information harvested from the physical implementation of the cipher such as timing analysis [8], power monitoring [9], fault attack [4], electromagnetic radiation [10], etc. Figure 5.3 shows a cryptographic model developed for side channel attacks [13].

Among all types of side channel attacks, power analysis is the most powerful attack since the power consumption of a cipher may provide a lot of information about the running operations and their involved parameters. Besides, this attack needs a very simple set of equipment such a PC with an oscilloscope and a small resistor in the power supply line to measure the power. Power analysis attacks have been proven to be very effective attacks against the implementation of many symmetric and public key algorithms [12].

In the RBS algorithm, the value of the NFSR and Accumulator registers are XOR-ed and stored in the Accumulator if the input message to the MAC is one unless nothing happens. Therefore, when the input is one, the power consumption

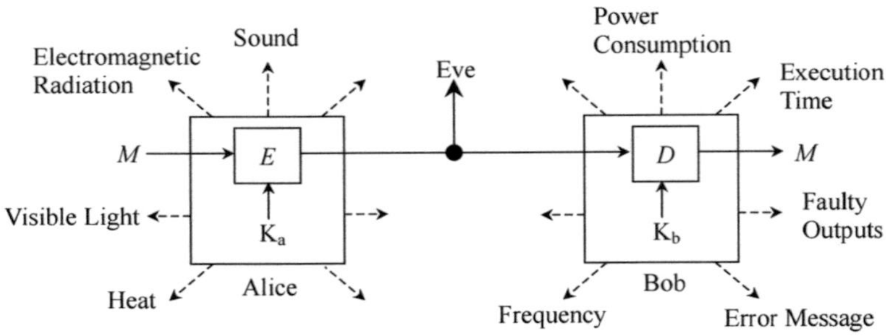

Fig. 5.3 Cryptographic model including side channel attacks presented in [13]

Fig. 5.4 Adding redundant MAC generator to RBS cipher

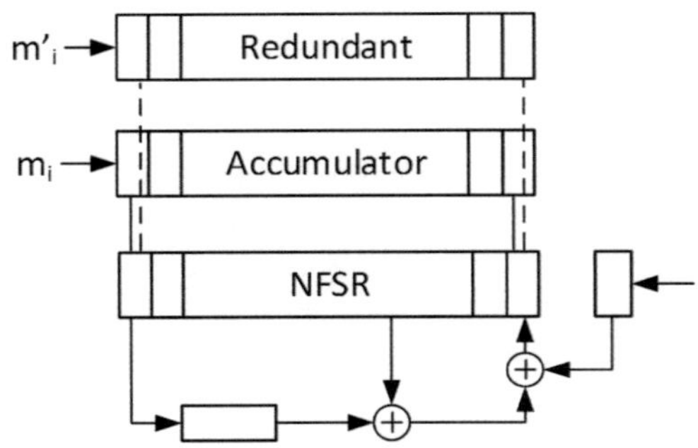

will be high to perform bitwise XORing of the two registers while the power consumption will be considerably low when the input is zero. However, the current will not be zero since some other operations such as calculating the nonlinear function of the NFSR and shifting it are still required to be performed. Based on this information, an attacker can find if the input message is zero or not during the generation of the redundant data by tracking the drained current in time. Also, the attacker can find the redundant data during generating the keystream since the redundant data is the input of the MAC in this process. By having the redundant data, the key space will be shrunk dramatically and finding the secret key for the attacker will be easier.

One solution for RBS to be resilient against side channel attacks is through adding extra hardware modules as a redundant circuit such that when the input message into the MAC is zero, this part will become activated such that the power consumption of the circuit goes higher. This trick can confuse the attacker in finding the value of input (Fig. 5.4). However, it is estimated that this solution will increase the cost in terms of the area and power consumption.

Another reasonable solution is adding extra hardware to the implementation in order to have a parallel architecture which can process two or more bits of input message at the same time (Fig. 5.5). This solution also results in increasing the cost of power and area by approximately 30–40 %. However, the resulting performance of the cipher will be improved 2 or 3 times.

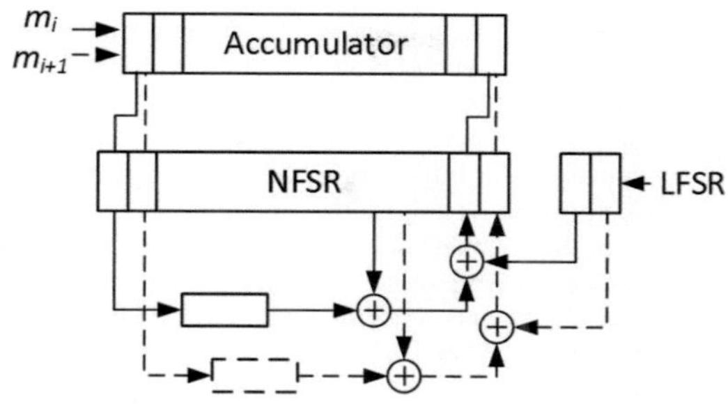

Fig. 5.5 RBS cipher with radix-2

5.4 Conclusion

In this chapter, the security of the RBS algorithm has been investigated against several existing attacks including the known-plaintext attack, chosen-plaintext attack, chosen-ciphertext attack, differential attack, related-key attack, substitution attack, linear attack, algebraic attack, cube attack and side channel attack. These attacks are very powerful attacks that have broken many contemporary ciphers. In this analysis, it was shown that the RBS cipher is or can be resilient against these attacks despite being a lightweight cipher. Future improvement of the considered attacks or composite attacks that incorporate several of these attacks can be analyzed based on the basic analysis presented in this chapter.

References

1. Agren, M., Hell, M., Johansson, T.: On hardware-oriented message authentication with applications towards RFID. In: Proceedings of International Workshop on Lightweight Security & Privacy (LightSec) (2011)
2. Anderson, R.: Security Engineering: A Guide to Building Dependable Distributed Systems, 2nd edn. Wiley, Indianapolis, Indiana (2008)
3. Biham, E.: New types of cryptanalytic attacks using related keys. IEEE Trans. Comput. **7**(4), Indianapolis, Indiana 229–246 (1994)
4. Boneh, D., DeMillo, R., Lipton, R.: On the importance of checking cryptographic protocols for faults. In: Advances in Cryptology - EUROCRYPT'97. Lecture Notes in Computer Science, vol. 1233, pp. 37–51. Springer, Berlin, Heidelberg (1997)
5. Courtois, N., Meier, W.: Algebraic attacks on stream ciphers with linear feedback. In: Advances in Cryptology – EUROCRYPT 2003. Lecture Notes in Computer Science, vol. 2656, pp. 345–359. Springer, Berlin, Heidelberg (2003)
6. Dinur, I., Shamir, A.: Cube attacks on tweakable black box polynomials. IACR Cryptology ePrint Archive. http://eprint.iacr.org/2008/385 (2008)
7. Dinur, I., Shamir, A.: Cube attacks on tweakable black box polynomials. In: Advances in Cryptology - EUROCRYPT 2009. Lecture Notes in Computer Science, vol. 5479, pp. 278–299. Springer, Berlin, Heidelberg (2009)
8. Kocher, P.: Timing attacks on implementations of Diffie-Hellman, RSA, DSS, and other systems. In: Advances in Cryptology - CRYPTO'96. Lecture Notes in Computer Science, vol. 1109, pp. 104–113. Springer, Berlin, Heidelberg (1996)

9. Kocher, P., Jaffe, J., Jun, B.: Differential power analysis. In: Advances in Cryptology - CRYPTO'99. Lecture Notes in Computer Science, vol. 1666, pp. 388–397. Springer, Berlin, Heidelberg (1999)
10. Kuhn, M., Anderson, R.: Soft tempest: hidden data transmission using electromagnetic emanations. In: Information Hiding. Lecture Notes in Computer Science, vol. 1525, pp. 124–142. Springer, Berlin, Heidelberg (1998)
11. Matsui, M., Yamagishi, A.: A new method for known plaintext attack of FEAL cipher. In: Advances in Cryptology – EUROCRYPT'92. Lecture Notes in Computer Science, vol. 658, pp. 81–91. Springer, Berlin, Heidelberg (1993)
12. Messerges, T., Dabbish, E., Sloan, R.: Examining smart-card security under the threat of power analysis attacks. IEEE Trans. Comput. **51**(5), 541–552 (2002)
13. Zhou, Y., Feng, D.: Side-channel attacks: ten years after its publication and the impacts on cryptographic module security testing (2005)

Chapter 6
RBS Performance Evaluation

Abstract This chapter presents a comprehensive evaluation of the proposed RBS cryptosystem and compares its performance against the performance of existing lightweight ciphers. An extensive set of simulation experiments is performed to show that the RBS cryptosystem requires less power and area compared to other known symmetric algorithms proposed for RFID systems especially when authentication is required. Such saving in the area overhead has a direct effect on the implementation cost of RFID tags which is another main practical concern. Our experiments also examine other performance metrics such as the energy-per-bit, the hardware efficiency, the area-time product and the power-area-time product. Simulation results demonstrate RBS superiority compared to existing algorithms. The chapter shows that the RBS algorithm is a promising candidate for providing strong resilience to several RFID security attacks—despite the severe resource constraints of passive RFID tags.

In Chap. 4, the RBS algorithm along with its proposed hardware implementation was introduced. Then, the security of the RBS algorithm against existing powerful and well known attacks was proven in Chap. 5. In this chapter, we present the experimental results of the Application-Specific Integrated Circuit (ASIC) hardware implementation of the RBS algorithm. We evaluate of the proposed RBS cryptosystem and compare its performance against the performance of existing lightweight block ciphers and stream ciphers such PRESENT [1], Trivium [5], Grain [7] and Hummingbird (HB-2) [3] discussed in Chap. 3. In this performance evaluation, AES [6] is used as a benchmark. We consider existing ciphers which provide message authentication as well as ciphers which do not provide such a service. The used performance metrics include the operating clock frequency, key size, size of the data block and the used technology to compare one-dimensional metrics such as area, performance, throughput, power and multi-dimensional metrics such as energy, hardware efficiency, area-time product and power-area-time product.

A. Khattab et al., *RFID Security*, Analog Circuits and Signal Processing,
DOI 10.1007/978-3-319-47545-5_6

6.1 ASIC Implementation of RBS

The different components of the RBS cryptophytes presented in Chap. 4 are implemented using Verilog hardware description language and synthesized via Synopsys Design Compiler in the 90 nm technology. It is worth mentioning that no tool optimization is set neither for area nor for power. The operating clock frequency is set to be 10 MHz. The operating conditions are set to be typical while the supply voltage is fixed at 1 V, and the temperature is set to 25 °C. Tables 6.1, 6.2 and 6.3 summarize the reported area and power for the different components of the RBS design, respectively.

The area report shown in Table 6.1 is based on the Gate Equivalence (GE) which is independent of the used fabrication technology. Therefore, changing the underlying technology does not have any effect on the area result which makes it a fair parameter in comparisons.

Power consumption in CMOS circuits consists primarily of a static power component dissipated by the leakage currents even when there is no change in the inputs, and a dynamic power component which is in turn comprised of the switching power consumed for charging and discharging load capacitances. Static power is a type of dissipation which is not affected by changes in the level of inputs and outputs. Current leakage is a technology dependent parameter. Therefore, static power dissipation differs in different technologies for the same design. In new technologies, the length of the channel becomes shorter which leads to smaller gate oxide thickness. Thus, newer devices have higher static power consumption. To have a fair comparison, all designs need to be fabricated in the same technology.

Similar to static power dissipation, dynamic power is also dependent on the underlying technology except that in a newer technology the dynamic power becomes less for the same design. This is because the size of the capacitance is

Table 6.1 Area of each component of RBS design [GE]

	RBS-83	RBS-100	RBS-116	RBS-132
MAC generation	678	816	905	1051
Enc/Dec cipher	10	10	10	10
Counter	53	53	53	73
Transmitter/Receiver	956	956	956	956

Table 6.2 Static power consumption of each component of the RBS design [μW]

	RBS-83	RBS-100	RBS-116	RBS-132
MAC generation	3.7953	4.5808	4.8024	5.89
Enc/Dec cipher	0.123	0.123	0.123	0.123
Counter	0.287	0.287	0.287	0.335
Transmitter/Receiver	4.53	4.53	4.53	4.53

Table 6.3 Dynamic power consumption for each component of different RBS designs [μW]

	RBS-83	RBS-100	RBS-116	RBS-132
MAC generation	15.6047	18.7028	23.1774	24.28
Enc/Dec cipher	0.143	0.143	0.143	0.143
Counter	0.38	0.38	0.38	0.414
Transmitter/Receiver	24.37	24.37	24.37	24.37

Table 6.4 Total area and power consumption overhead for different RBS designs

	RBS-83	RBS-100	RBS-116	RBS-132
Area [GE]	688	826	915	1061
Power consumption [μW]	19.67	23.55	28.25	30.46

reduced and the power needed to charge and discharge them becomes less too. Besides, dynamic power is dependent on the switching of the logics at their outputs too. The number of switching in a circuit increases proportionally with the operating frequency. Therefore, in the comparison of the dynamic power of different designs, it is required for all designs to work at the same frequency and have the same technology. Table 6.3 demonstrates the dynamic power dissipation for different modules in different RBS designs.

Among the listed modules in Tables 6.1, 6.2 and 6.3, the counter and the transmitter/receiver are common parts of every typical data communication system regardless of whether the data must be encrypted/decrypted or not. Therefore, these parts are not considered as overhead in the RBS implementation. In other words, the RBS algorithm adds only the MAC generator and the Enc/Dec Cipher modules to the system. The total power consumption is obtained from the summation of the static power and dynamic power. Considering just RBS modules, the total area and also total power consumption overheads of RBS implementation is calculated in Table 6.4.

Concerning the required clock cycles, encryption/decryption in the RBS algorithm is performed along with data transmission/reception and the performance of RBS is only limited by the time required for generating the MAC outputs which are the redundant bits and the keystream. Generating the redundant bits takes $m + 1$ clock cycles, where m is the length of the plaintext. Producing the keystream takes n clock cycles, where n is the size of the redundant bits. Besides, one clock cycle is required for the bitwise addition of the keystream and the plaintext plus 2 cycles for generating the authentication keys. Altogether, $m + n + 4$ clock cycles is the overhead for encryption/decryption plus authentication in the RBS algorithm. For example, in RBS-132, 65 clock cycles are needed for generating the redundant data and 68 clock cycles for the keystream. Table 6.5 demonstrates the total number of clock cycles to generate the ciphertext for different designs of the RBS algorithm.

Table 6.5 The number of clock cycles required for generating the output in RBS

Cipher	Redundant bits time overhead	Keystream time overhead	Total
RBS-83	41	43	87
RBS-100	49	52	104
RBS-116	53	56	120
RBS-132	65	68	136

6.2 Comparison of Ciphers

This section presents a comprehensive analysis of all ciphers introduced in Chap. 3 from a hardware perspective. The number of ciphers designed and published recently for restricted resources is more than those investigated in this book. However, only those ciphers that cryptanalysis indicates promising security, and also considered to have low resource hardware, have been chosen to be included in our comparative study.

For comparison, considering only one metric such as the speed or the area results into a one-dimensional analysis which is not effective in finding the best cipher since each cipher might be good at just one of the metrics while in other metrics they might be unacceptable. In contrary, a cipher might be worse in one of metrics while in other metrics it might be better than other competitors. To have a comprehensive analysis, the hardware performance is required to be investigated in multiple dimensions with various quantities such as area, performance, throughput, power, and energy. The AES algorithm is used as the benchmark for comparison. In order for a cipher to be counted as lightweight cipher, it should be smaller and faster than AES.

The RBS algorithm is compared with five other encryption algorithms in terms of (1) the required key and initial vector size, (2) the data block size, (3) the required number of clock cycles for completing the encryption process, (4) the two input NAND GE equivalent area, and (5) the total power consumption when the clock frequency is set to 10 MHz. Table 6.6 summarizes the comparison results in terms of area, performance and power. In this table, the initial clock cycles are shown in parenthesis.

The reports of all the compared methods are for 130 nm technology, except PRESENT which use 180 nm. Whereas RBS is synthesized in the newer 90 nm technology which has considerable effect in area and power overhead. However, area reports are given in Gate Equivalent (GE) which is independent of the used technology. Considering the effect of technology scaling (α) on power consumption, the total power could be normalized by α^2.

The timing, area, and power consumption reports in Table 6.6 for AES, PRESENT, Trivium, Grain, and HB-2 algorithms are calculated without considering the authentication's implementation. Meanwhile, the RBS algorithm provides also authentication service as mentioned before. Besides, the area and power reports for RBS algorithm listed in Table 6.6 include the MAC generator part's area/power

Table 6.6 Comparing RBS with other encryption methods

	Key/IV [bits]	Block size [bits]	Clock cycles	Area [GE]	Total power [μW]	Freq. [MHz]	Tech. [nm]
AES	128/–	128	160	3200	300	10	130
PRESENT	128/–	64	32	1884	7.34	0.1	180
Trivium	80/(80)	1	(1333)[a]+1	2599	181.18	10	130
Grain	128/(128)	1	(512)[a]+1	1857	167.73	10	130
HB-2	128/64	16	(80)[a]+16	2332	156.8	10	130
RBS-83	83/–	40	(86)[a]+87	688	19.67	10	90
RBS-100	100/–	48	(104)[a]+104	826	23.55	10	90
RBS-116	116/–	56	(120)[a]+120	951	28.25	10	90
RBS-132	132/–	64	(68)[a]+136	1061	30.46	10	90

[a] Cycles required for initialization

overheads which are still better than other compared ciphers. Similar to the area and power overheads, the RBS timing overhead is still comparable with other algorithms while their reported timing overhead just considers the encryption/decryption process.

Comparing all metrics in one table gives a lot of information which makes finding the best cipher very difficult. In the following subsections, ciphers are compared for only one metric then they are studied under two or more different metrics in two states. First, when all of the competitor ciphers provide only confidentiality. Second, when all ciphers provide message authentication service along with confidentiality.

Providing the authentication service in ciphers—especially stream ciphers like Grain and Trivium—is necessary since the ciphertext is the result of the bitwise addition of the plaintext and the keystream. Therefore, if the attacker knows the plaintext by changing some bits of the ciphertext, it can easily replace its own message with the real message while the receiver accepts the message since it does not have any knowledge that the message has been manipulated. This fact demonstrates the importance of providing authentication in stream ciphers. On contrary, changing any bits of the ciphertext in block ciphers, the decryption party may generate an irrelevant and meaningless message from the corresponding block. However, providing the authentication in block ciphers is still necessary since it plays a basic role in providing privacy.

Generally speaking, current hash functions are not acceptable for performing authentication in constrained environments. They require significant amounts of overhead and they are not hardware friendly. Table 6.7 shows the results of the hardware implementation of some MAC algorithms designed for RFID systems. As this table states, either their area or their performance overhead is still very high to be integrated with a cryptosystem. Therefore, in case of comparison of authenticated ciphers, those ciphers that cannot provide authentication alone will not be considered.

Table 6.7 Hardware implementation of MAC

Design	Area [GE]	# of cycles	Dynamic power	Leakage power	Freq. [Hz]	Tech. [nm]	Output size [bits]
PSQUASH [11]	1624	25.1 K	18.3 nW	7.7 nW	100 K	130	48
SHA-1 [9]	4276	405	3.74 μW	23 μW	500 K	130	160
SHA-256 [4]	10,868	1128	52.37 μW	N/A	100 K	350	224
MD4 [4]	7350	456	N/A	N/A	N/A	130	128
MD5 [10]	10,332	68	N/A	N/A	133 M	130	128
RIPEMD [10]	17,446	96	N/A	N/A	143 M	130	160

Table 6.8 The performance of different hash functions based on PRESENT

	Hash output size [bits]	Input size [bits]	Cycles per block	Area [GE]	Power [μW]
DM-PRESENT-80	64	80	33	2213	6.28
DM-PRESENT-128	64	128	33	2530	7.49
H-PRESENT-128	128	64	32	4256	8.09
C-PRESENT-192	192	64	108	8048	9.31

Apart from the RBS algorithm which provides a message authentication service along with providing confidentiality, other ciphers either provide this service optionally by applying some small modifications to their hardware such as HB-2 and Grain, or require to be integrated with other algorithms to provide this service as in Trivium and AES. Also, there are some ciphers which have their own dedicated hash functions for this purpose like PRESENT presented in [2]. To compare ciphers when they provide authentication, those which have the ability to authenticate the message will be considered like RBS, HB-2, PRESENT and Grain. Since there is no dedicated hardware for authentication for Trivium and AES, these two ciphers are not considered in the comparison of ciphers when the authentication service is provided.

Grain and HB-2 use the same hardware used for confidentiality for authentication with some modifications. Unfortunately, there is no official report for its area and power overheads. Therefore, in this comparison, it will be supposed that this overhead is negligible. However, since the confidentiality and authentication services cannot be performed at the same time because of sharing the hardware resources, the time required to produce the output will be the summation of time needed for performing for each of them. For PRESENT, some dedicated hash functions are introduced in [2]. Their output sizes are fixed to 64, 128 and 192 bits while their inputs are 64, 80 and 128 bits (Table 6.8). Among these designs, for the sake of collision attack and also satisfying conditions of comparison, H-PRESENT-128 is considered for comparison of authenticated ciphers which uses PRESENT core and supports 64-bit input and 128-bit output at 180 nm and 100 kHz frequency. Likewise, encryption and authentication do not happen at the same time for this cipher too because of resource sharing as the case with Grain and HB-2.

For authentication, HB-2 adds a payload of 64 bits for messages between one to eight words. Meanwhile, the authentication code in Grain is at least 32 bits for messages smaller than 32 bits. For messages longer than 32 bits, it is required to expand the authentication message with the message to prevent substitution attack. To have a fair comparison with ciphers providing the authentication service, it is supposed that the length of plaintext input for all ciphers is 64 bits.

6.2.1 Area

The implementation area is one of the important parameters for comparing different algorithms. It states the amount of silicon required for the core design, excluding power rings and input/outpot (I/O) cells. This metric is typically expressed in μm^2. However, the more practical independent method of expressing the area is to express it in terms of the Gate Equivalence (GE), calculated by dividing the total area by the lowest power two-input NAND gate's area. In definition of the lightweight encryption algorithm, a complete RFID tag, including the analog part, might have between 1000–10,000 GE, and for the security module this margin may be kept between 200–2000 GE [8]. Based on this defined limitation, having lower area for hardware implementation is one of the most important factors to help finding a better cipher for encryption. In this metric, the one has the lower area is the better one.

Figure 6.1 compares all Elliptic Curve Cryptography (ECC) designs designated for restricted resource environments introduced in Chap. 3. This figure shows that the total area for these designs ranges between 30–6 k GE which is 15–3 times more than the area limitation defined for lightweight cryptosystems while these designs are limited to a special prime fields or one special elliptic curve. The result in this figure explains why public key encryption algorithms are not proper for providing security in RFID systems.

Relying on just one metric like the area to remove a cipher from the list of the lightweight cryptosystems is not acceptable. However, regarding to the reported performance for ECC designs in Chap. 3, public key cryptosystems are several times slower than private key cryptosystems in terms of performance. This correlates at least with a two to three orders-of-magnitude higher power consumption. Also the time for key computation is another overhead which is not a negligible factor in asymmetric algorithms.

These two metrics, area and performance, are enough to take asymmetric algorithms out of the comparing ciphers list despite all of the advantages of these algorithms over symmetric algorithms like key exchange and key management which they provide. From now on, just private key ciphers will be considered for comparison based on the different metrics.

Figure 6.2 compares ciphers based on the area metric for three block ciphers, AES, PRESENT and RBS, two stream ciphers, Grain and Trivium and one hybrid cipher, HB-2 while AES is considered as a benchmark in this comparison. It must be noticed that the hardware implementation for AES and PRESENT in this report is

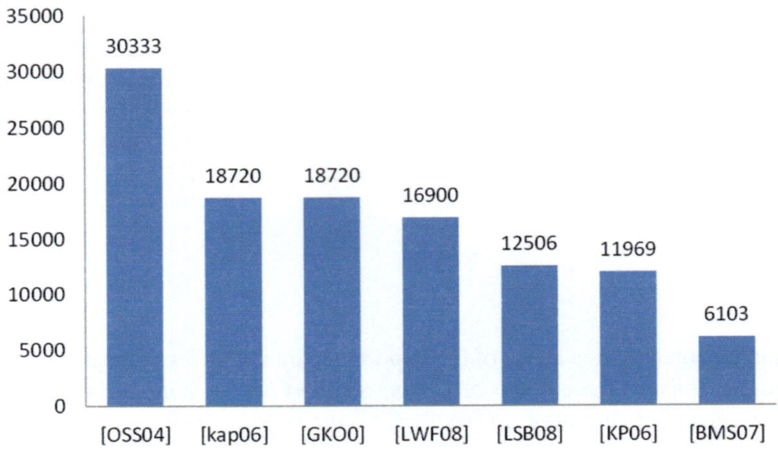

Fig. 6.1 Comparing the area of different ECC designs

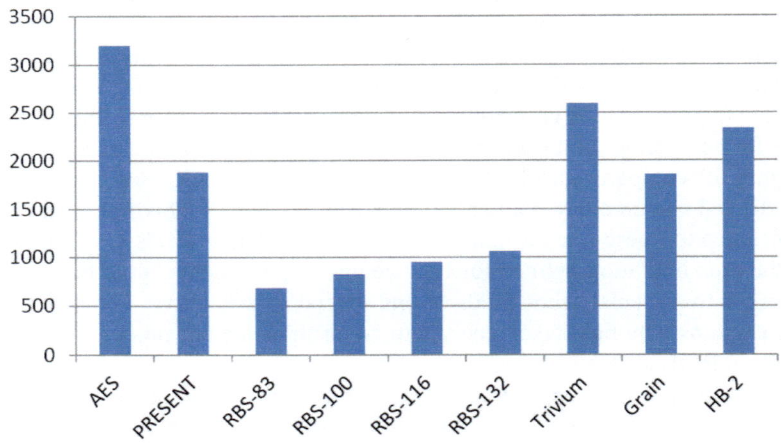

Fig. 6.2 Area comparison of symmetric ciphers without providing authentication

only provided for the encryption process. In order to support the decryption process, extra hardware is required which makes their total area more than the reported one. On the contrary, Trivium and Grain support both the encryption and decryption processes since both processes need the same hardware. However, the area report of Trivium does not cover the authentication part which implementation will impose extra area. To add authentication service to Grain and HB-2, some modifications to the presented hardware are needed to use it for this purpose too which means a small overhead in area.

Regarding Fig. 6.2, RBS designs have the smallest area overhead among all of the ciphers. The area of RBS-132 is about three times less than area of AES. After RBS,

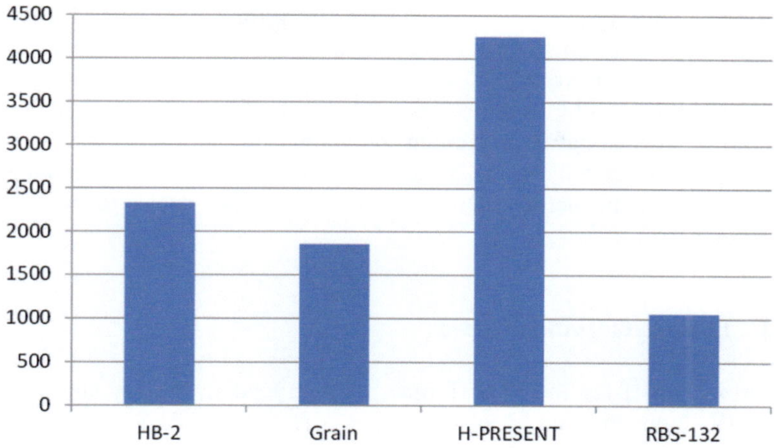

Fig. 6.3 Area comparison of different ciphers with providing authentication

Grain has the smallest area while the difference between these two ciphers is 796 GE which means RBS-132 is 43 % smaller than Grain. This shows the significant advantage of RBS over other ciphers in terms of the area metric. Among the four different RBS designs, RBS-83 has the smallest area overhead. RBS-83 has 1.5 times less area than RBS-132. This result is predictable since the length of the supporting plaintext and the secret key for this design is the least among all of RBS designs. However, this benefit in area overhead is obtained at the cost of degrading the security strength.

Figure 6.3 compares four ciphers when they provide authentication besides confidentiality. Among these ciphers, RBS, Grain and HB-2 are using the same hardware to provide authentication while PRESENT in [2] have an overhead in area to adapt the use of a hash function with its cipher. Again, RBS has the smallest area among all of these ciphers.

6.2.2 Performance

Performance is one of the important metrics which helps to compare different designs in term of speed. This metric can be studied in two aspects. First, the total number of clock cycles required for generating the ciphertext in the encryption process by the ciphers. Since it is independent of the used fabrication technology and the operating clock frequency, it will be a suitable criterion for the comparison of different ciphers. This metric will be helpful in finding the number of bits completed per cycle (bits-per-cycle) and the consumed energy in the ciphers. The first aspect of performance, the total number of clock cycle required for completing processes, is stated in two parts: load or initialization cycles and computation cycles.

The second aspect of the performance metric is the time in seconds which it taken to complete a process. This aspect is dependent on the maximum frequency that the designed hardware can work in. The maximum frequency of a circuit is defined by the worst delay caused by its critical path. Measuring the performance in time will be helpful in calculating the maximum throughput. In the following subsections, each of these aspects will be explained and symmetric ciphers will be compared based on them. In this metric, the one with the lower clock cycles for computation is the better cipher.

6.2.2.1 Load/Initialization Cycles

This period starts from the RESET time, through loading the key and the initial vectors (IV), until the first bit of the output is ready. The RBS cipher is similar to the Trivium, Grain, and HB-2 ciphers in using initial vectors (IV) for refreshing the key which imposes extra clock cycles for initializing the cipher process during the algorithm startup or whenever the key changes. The number of cycles for initialization is independent of the size of the data block and varies in different designs.

In Table 6.6, the number of clock cycles required for initialization for each cipher is depicted in parenthesis. For RBS designs, the key initialization happens once every 132 clock cycles. The result is kept as the authentication key and bitwise addition of this key with the input will be applied to the MAC generator after the process of initialization of the MAC. Since initialization happens once for several messages, this overhead is not considered for performance comparison. However, these initial vectors for stream ciphers open new opportunities for attackers since the generated key is a function of initial vectors.

6.2.2.2 Computation Cycles

The second part of the performance metric is the computation cycles which states the number of cycles required to encrypt the message. Despite the initialization, this one depends on the length of the input message. Table 6.9 displays the number of cycles required for computation for different sizes of data blocks in different cryptosystems.

For HB-2, Grain and Trivium, the number of clock cycles increases with increasing the size of the data block. While in PRESENT, for data blocks equal to or smaller than 64 bits, fixed clock cycles are needed since the number of rounds, the size of plaintext and ciphertext in this cryptosystem are fixed to 32, 64, and 64 respectively. In AES, the number of clock cycles for data blocks is equal to or smaller than 128 bits is fixed to 160 cycles as the case in PRESENT.

In RBS, the number of clock cycles for the same size of data blocks will change, depending on the chosen design, since each design works on a specified fixed size of message. For instance, to encrypt a 48-bit message, in RBS-83 it takes 174 cycles

Table 6.9 Comparison of clock cycles to encrypt a message

Cipher	Init	16 bits	32 bits	48 bits	64 bits	96 bits	128 bits
HB-2	16	16	32	48	64	96	128
Grain-128	513	16	32	48	64	96	128
Trivium	1333	16	32	48	64	96	128
PRESENT-80	0	32	32	32	32	64	64
PRESENT-128	0	32	32	32	32	64	64
AES-128	0	160	160	160	160	160	160
RBS-83	43	87	87	174	174	261	348
RBS-100	52	104	104	104	208	208	312
RBS-116	60	120	120	120	240	240	360
RBS-132	68	136	136	136	136	272	272

Table 6.10 Number of required cycles for encrypting 64-bit plaintext plus authentication

	Encryption [cycles]	Authentication [cycles]	Total [cycles]
HB-2	64	64	128
Grain-128	64	64	128
H-PRESENT	32	32	64
RBS-132	64	68	132

while in RBS-100 it takes 104 cycles. Thus, it is required to choose the right design based on the size of the message before implementing it. For messages equal to or shorter than 40, 48, 56 and 64 bits, it would be better to use RBS-83, RBS-100, RBS-116 and RBS-132, respectively, in terms of timing and size of ciphertext.

Comparing all the designs in Table 6.9, RBS has the highest number of clock cycles for each size of data block. However, in this comparison, it must be stated that except RBS, all other designs do not compute the authentication code. Table 6.10 shows the total number of cycles required to produce the ciphertext plus the authentication code for different cryptosystems when the size of the plaintext is 64 bits.

Comparing the results in Table 6.10, PRESENT has the lowest number of cycles while RBS needs the highest number of cycles to complete its process. However, RBS-132 needs only 8 clock cycles more than HB-2, Grain which means 6 % more clock cycles. This difference results from the difference between the size of MACs and the extra operations for preparing the authentication keys.

6.2.2.3 Bits-per-Cycle

One of the helpful metrics in evaluating the performance of a cryptosystem is the bits-per-cycle derived from the computation time. For stream ciphers, the bits-per-cycle means the number of bits in the output keystream per clock cycle. However, this definition can be expanded for all ciphers to describe their output rate.

Table 6.11 Bits-per-clock without authentication

Cipher	Block size [bits]	Output size [bits]	Clock cycles	Bits per cycle
HB-2	16	16	16	1
Grain-128	1	1	1	1
Trivium	1	1	1	1
PRESENT-80	64	64	32	2
PRESENT-128	64	64	32	2
AES-128	128	128	160	0.8
RBS-83	40	83	87	0.46
RBS-100	48	100	104	0.46
RBS-116	56	116	120	0.47
RBS-132	64	132	136	0.47

Table 6.12 Bits-per-clock with authentication

Cipher	Plaintext [bits]	# of clocks for enc.	MAC [bits]	# of clocks for MAC	Bits per clock
HB-2	64	64	64	64	1
Grain-128	64	64	64	64	1
H-PRESENT	64	32	128	32	3
RBS-132	64	66	68	70	0.97

Therefore, this definition is modified to the number of bits in the output divided by the number of cycles per block to cover block ciphers and hybrid ciphers [5]. According to this metric, the cipher that has the higher bits-per-cycle is the better one.

Based on the bits-per-cycle metric, Table 6.11 compares the number of bits, encrypted with different ciphers for various amounts of data when authentication is not included. Among all ciphers in this table, AES has the lowest rate of output while PRESENT has the highest one. Since none of the competitors in Table 6.11 provide MAC in their output, the RBS redundant part (MAC) is not considered as a part of the output and it is supposed that the length of the new output is equal to the length of the generated keystream not the ciphertext.

Table 6.12 compares the bits-per-clock while encryption is performed along with authentication. To have a fair comparison, it is supposed that all ciphers use 64-bit data block to encrypt. For Grain and HB-2 ciphers, the rate of the number of produced output bits to the number of required clock cycles for generating them is the same when authentication is included and when it is not. In this comparison, RBS cipher is very close to HB-2 and Grain ciphers.

Table 6.13 Maximum clock frequency

Cipher	Tech. [nm]	Maximum freq. [MHz]	Normalized freq. [MHz]
HB-2	130	N/A	N/A
Grain-128	130	925.9	925.9
Trivium	130	358.4	358.4
PRESENT-128	180	N/A	N/A
AES-128	130	130	130
RBS	90	5000	2500

6.2.2.4 Maximum Clock Frequency

The connections between the inputs/outputs and the registers form timing paths between them. Among these paths, the slowest path in the design is known as the critical path which defines the upper bound on the clock frequency. The operating clock frequency of a design is usually at a significantly lower rate than the maximum frequency. In this metric, the cipher with a higher maximum frequency is the better one.

The maximum frequency for each cipher is given in Table 6.13 along with the used technology. For HB-2 and PRESENT, no maximum frequency for the ASIC implementation has been reported.

To find the maximum frequency for all RBS designs, the operating frequency in simulation was increased to 5 GHz while the time slack was still positive. The maximum frequency depends on the used technology. In newer technologies, the size of the capacitance loads is scaled down. Therefore, the circuit can work at a higher frequency. To compare all ciphers for this metric, it is required that all ciphers are implemented with the same technology or the frequency become normalized with the technology scaling factor α.

In Table 6.13, after normalization, RBS ciphers still have the highest maximum frequency among all ciphers. The reason behind that is having a simple circuit with short paths in their implemented hardware. On contrary, AES has the lowest maximum frequency among the ciphers.

6.2.2.5 Throughput

Throughput is the rate of producing the new output with respect to time, typically expressed in bits-per-second [5]. This metric reaches its sustainable rate when the initialization is completed at a given operating clock frequency. Therefore, it is simply calculated through multiplying the bits-per-cycle by the clock frequency. To have a fair comparison, it is required that all competitors work in the same operating clock frequency. For this metric, the cipher that has higher throughput is the better cipher.

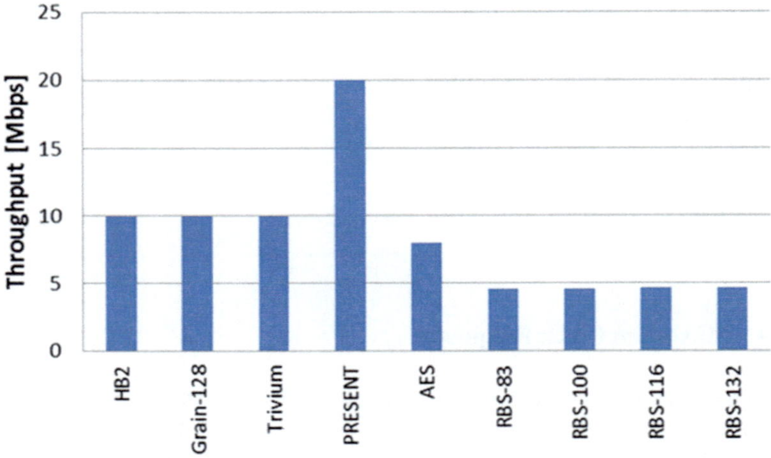

Fig. 6.4 Throughput when the operating frequency is 10 MHz without authentication

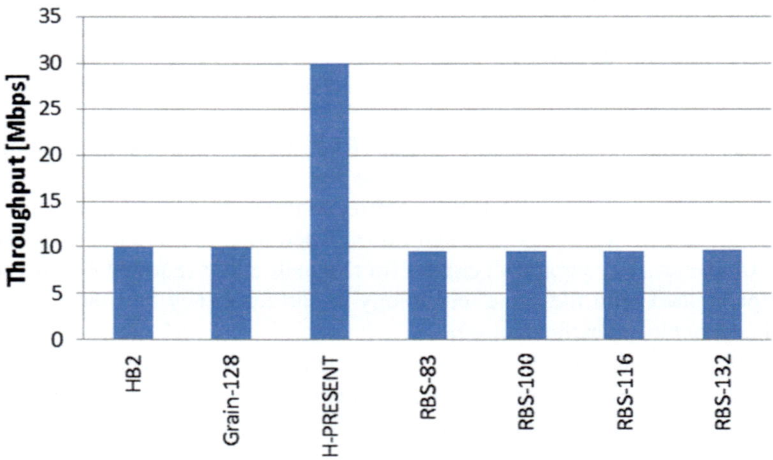

Fig. 6.5 Throughput when the operating frequency is 10 MHz with authentication

Figure 6.4 compares the throughput of all ciphers when the operating frequency for all ciphers is set to 10 MHz and ciphers do not provide the authentication service. In this figure, PRESENT has the highest throughput among all ciphers since it has the highest bits-per-cycle. On the contrary, RBS has the lowest throughput in this comparison because of its low bits-per-cycle.

Figure 6.5 compares the throughput of the different ciphers when the operating frequency is set to be 10 MHz when all ciphers provide the authentication service too. In this comparison, PRESENT still has the highest throughput. However, the throughput of RBS is very close to HB-2 and Grain ciphers.

Table 6.14 Maximum throughput

Cipher	Bits per cycle	Maximum frequency [MHz]	Maximum throughput [Mbps]
HB-2	1	N/A	N/A
Grain-128	1	925.9	925.9
Trivium	1	358.4	358.4
PRESENT-128	2	N/A	N/A
AES-128	0.8	130	104
RBS-83	0.46	2500	1150
RBS-100	0.46	2500	1150
RBS-116	0.47	2500	1175
RBS-132	0.47	2500	1175

Table 6.15 Energy required for the encryption of a 64-bit plaintext without authentication

Cipher	Normalized total power [μW]	Clock cycles	Energy [nJ]
HB-2	156.8	64	1.0035
Grain-128	167.73	64	1.0734
Trivium	181.18	64	1.1595
PRESENT-128	382.86	32	1.2251
AES-128	300	160	2.4
RBS-132	60.92	136	0.8285

The maximum throughput will occur at the maximum clock frequency. In this metric, the cipher that has the higher maximum throughput is the better cipher. Table 6.14 compares the maximum throughput of all ciphers. Despite the low bits-per-cycle, RBS ciphers have the highest maximum throughput because of having the highest maximum frequency among ciphers.

6.2.3 Area-Time Product

The area of the hardware implementation and the time performance are two important metrics which have been so far studied in this chapter separately. The area-time product is a cost function which is equal to the product of the time taken to produce each new output bit and the area of the design [5]. To have the optimal value for this metric, it is required that the implemented hardware has low overhead in area while it provides high speed in producing the output. This metric is expressed in gate equivalent by second [GE-μs]. In this metric, the cipher that has the lower product is the better one. Table 6.15 demonstrates this metric for each cipher when the operating frequency is set to be 10 MHz and the ciphers do not provide the authentication service.

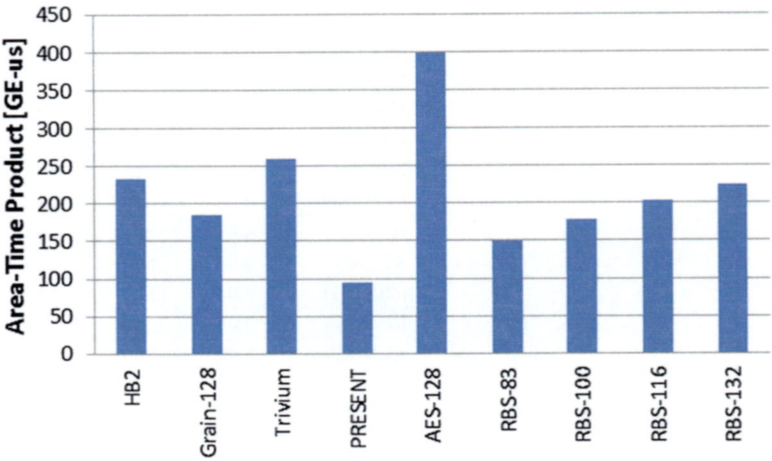

Fig. 6.6 Area-time product when authentication is not provided

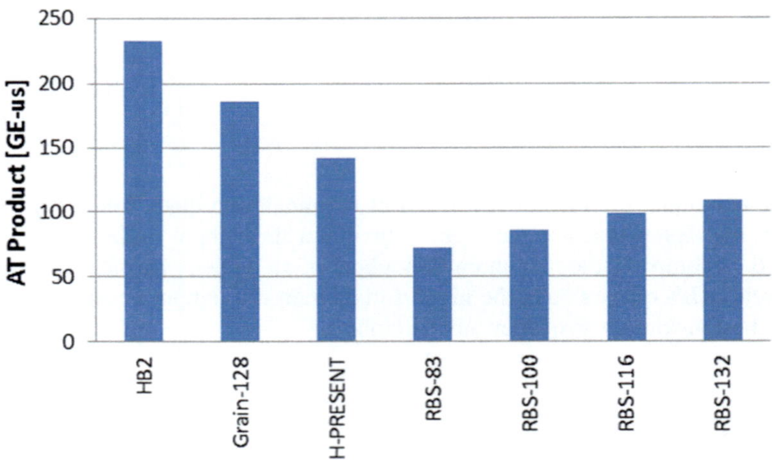

Fig. 6.7 Area-time product when authentication is provided

In Fig. 6.6, PRESENT has the least area-time product among all ciphers. After that, RBS and Grain ciphers have the next best area-time products.

Figure 6.7 shows the area-time product of the different ciphers when the operating frequency is 10 MHz and all ciphers provide the authentication service as well. Among all of these ciphers, RBS ciphers have the best (i.e., least) area-time product.

6.2.4 *Hardware Efficiency*

The total hardware performance cannot be determined by only measuring the gate count or time performance. The hardware efficiency is defined as the throughput per gate which states the balance between the size of the implemented hardware and its speed [5]. It is simply calculated by dividing the throughput by the area overhead. The hardware efficiency is expressed in Mbits per second per gate equivalents [Mbps/GE]. To have the highest hardware efficiency, it is required that an optimal balance exists between the throughput and the size of hardware. According to this metric, the cipher that has the higher efficiency is the better one.

Figure 6.8 demonstrates the hardware efficiency when the operating frequency is set to 10 MHz and the ciphers do not provide the authentication service. In this comparison, PRESENT has the highest hardware efficiency among all ciphers. After that, RBS designs and Grain has the best rate in hardware efficiency.

Figure 6.9 demonstrates the hardware efficiency when the operating frequency is 10 MHz and the different ciphers provide the authentication service. RBS-132 has the highest hardware efficiency among all ciphers.

6.2.5 *Power*

This metric demonstrates the required power for computation at each clock cycle. According to this metric, the cipher that needs a lower power is the better one. The total power consumption is obtained by adding up the static power and the dynamic power. To compare the total power dissipation in different ciphers, it is necessary that all designs have been fabricated at the same technology and also

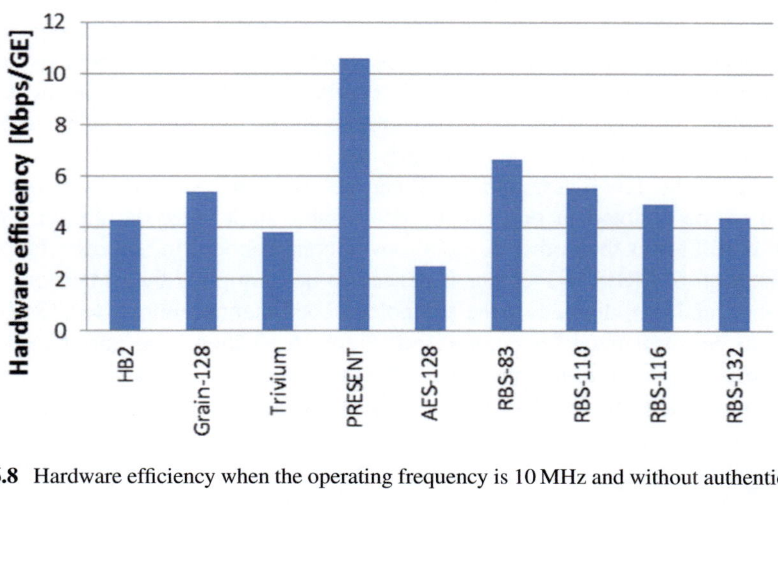

Fig. 6.8 Hardware efficiency when the operating frequency is 10 MHz and without authentication

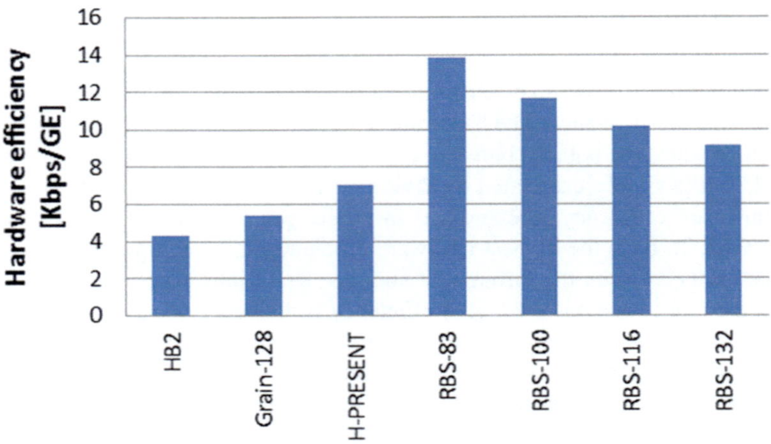

Fig. 6.9 Hardware efficiency when the operating frequency is 10 MHz and authentication is provided

work at the same operating frequency. Since both power components also depend on the supply voltage, the typical core voltage should be used. However, it is difficult to satisfy these conditions for all competitors. The measured power can be scaled with an acceptable margin of error to other frequencies and technologies if the static and dynamic components are treated separately. With this assumption, the dynamic power can be assumed to be directly proportional to the frequency and inversely proportional to the technology coefficient scaling, α.

At low frequencies, the static power is significant whereas at the other frequencies it may be trivial. Unfortunately, there is no straight formulation for normalizing the static power as the case with the dynamic power. However, these effects are assumed to be negligible and are not considered for comparison for different technologies since the static power represents a small percentage of the total power. Therefore, the total power of ciphers in different frequencies can be approximately normalized by multiplying it with the coefficient of frequency growth and multiplying by α^2 when the used technologies are different. It is not accurate but it gives a good estimation for comparison.

Based on the above assumptions, it can be estimated that the power report in RBS must be doubled in order to be comparable with other designs in 130 nm which is still lower than other designs' power consumption. In contrast, the power consumption of PRESENT cipher is needed to be multiplied by 10 because of its frequency, and then divided by the technology coefficient scaling, $\alpha = (180/130)^2$ or 1.9 to be normalized for comparison. Figure 6.10 shows the normalized total power consumption for all ciphers when the operating frequency is 10 MHz and the used fabrication technology is 130 nm. In this table, all RBS designs have the lowest power consumption among all ciphers.

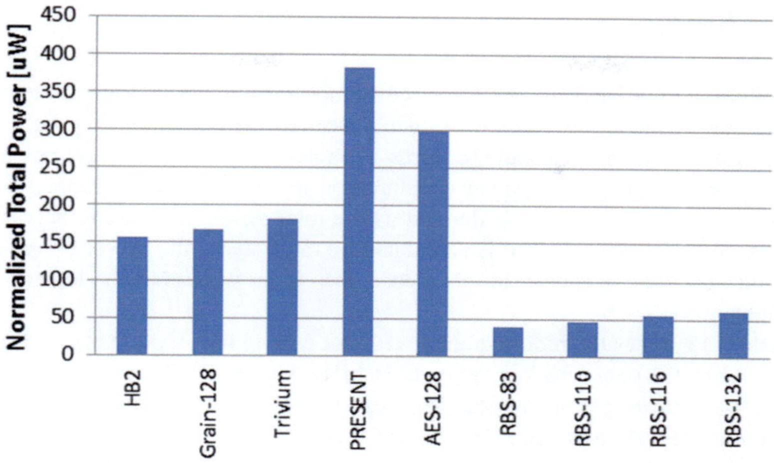

Fig. 6.10 Power consumption without authentication

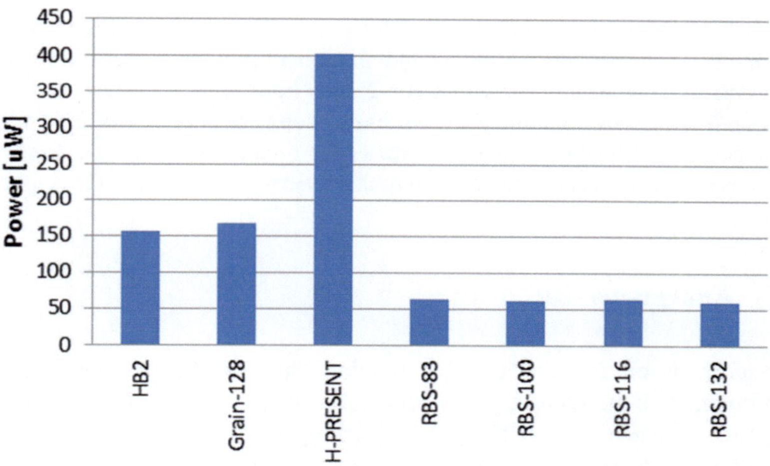

Fig. 6.11 Power consumption for 64-bit plaintext when authentication is provided

The normalized power consumption for ciphers at 130 nm and a frequency of 10 MHz when the authentication is provided is shown in Fig. 6.11. In this figure, the power consumption for Grain and HB-2 are the same as in the last figure since they use the same resource with no overhead in area. The power consumption of H-PRESENT is the average of the power consumption required for encryption (7.34 μW) and the power needed for authentication service (8.09 μW). The power consumptions of RBS-83, RBS-100 and RBS-116 ciphers are increased in this figure since these ciphers perform encryption and authentication for 64-bit plaintexts.

6.2.6 Energy

In battery operated devices, the power consumption may not be a good metric for comparison. Instead, the amount of energy needed for operation may be a more useful criterion because a battery stores a limited amount of energy, not power [5]. Two other metrics, the power consumption and the time performance are the basics of this metric. Energy is defined as the total power required to accomplish an operation in a given time. It is calculated by multiplying the power by the taken time and expressed in joules. In this metric, the cipher which has the lower energy is the better one.

Table 6.15 shows the required energy for encrypting 64-bit plaintext in different ciphers when the operating frequency is 10 MHz. Except RBS, none of the ciphers provide the authentication service. It should be noticed that since AES works on 128-bit data blocks, it is supposed that AES is encrypting two 64-bit plaintexts in order to have a fair comparison. In this comparison, RBS with 0.82 pJ needs the lowest energy while AES needs the highest energy which is three times more than RBS-132.

Table 6.16 shows the required energy for encrypting a 64-bit plaintext when the operating frequency is set to 10 MHz and all ciphers provide authentication beside confidentiality. The results show that HB-2 and Grain need 2.5 times more energy than RBS-132 to provide authentication. In this comparison, H-PRESENT uses the highest energy to encrypt and authenticate a 64-bit plaintext. On the contrary, RBS-132 needs the lowest energy to perform both services compared to other ciphers.

6.2.7 Energy-per-Bit

The energy-per-bit is calculated by dividing the total power consumption by the throughput. It is required to calculate this metric when both the power and throughput are measured at the same operating clock frequency. It seems that this metric is independent of frequency. However, the static power may dominate the total power consumption in low frequencies. Meanwhile, dynamic power has a substantial effect on it at higher frequencies [5]. According to this metric, the cipher which has the lower energy-per-bit is the better one. Figure 6.12 shows the energy-per-bit required for each cipher when the operating frequency is 10 MHz and

Table 6.16 Energy required for the encryption and authentication of a 64-bit plaintext

Cipher	Normalized total power [μW]	Clock cycles	Energy [pJ]
HB-2	156.8	128	200.7
Grain-128	167.73	128	214.68
H-PRESENT	402.42	64	257.5
RBS-132	60.92	136	82.85

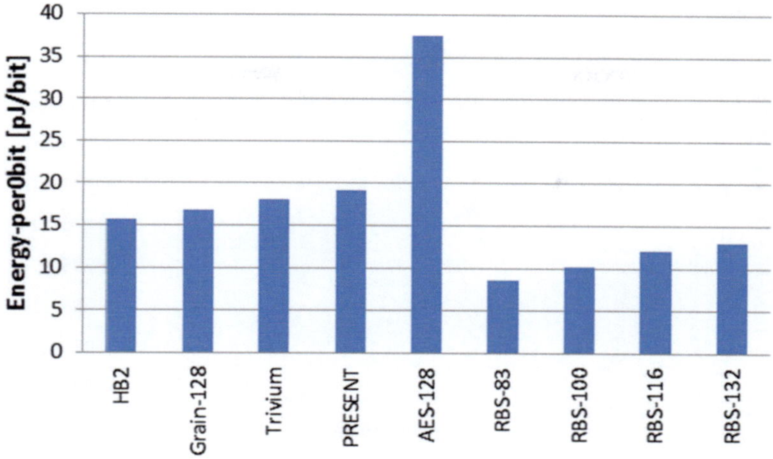

Fig. 6.12 Energy-per-bit without authentication

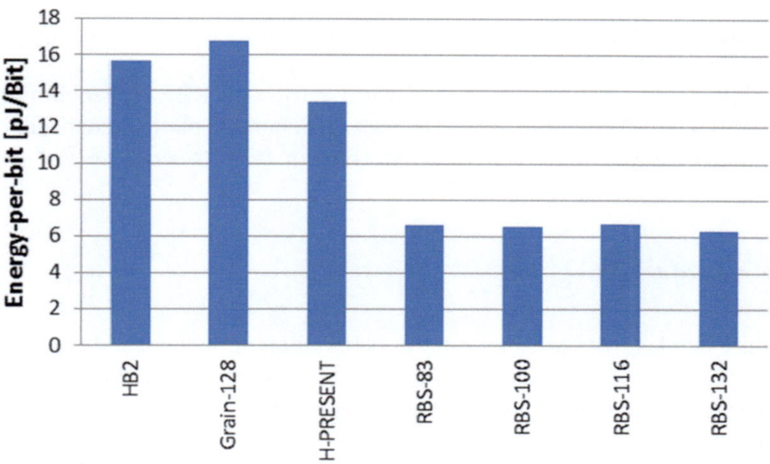

Fig. 6.13 Energy-per-bit with authentication

the used technology is 130 nm when all ciphers do not provide the authentication service. Since the measured power for all ciphers does not satisfy these conditions, the normalized power from Table 6.10 is used for this comparison.

In Fig. 6.12, RBS ciphers have the lowest energy-per-bit while it has the lowest throughput. On the contrary, PRESENT has the highest energy-per-bit after AES while PRESENT has the highest throughput among all ciphers.

Figure 6.13 shows the energy-per-bit required for each cipher when the operating frequency is 10 MHz and the used technology is 130 nm when all ciphers provide the authentication service. In this comparison, RBS ciphers again have the lowest

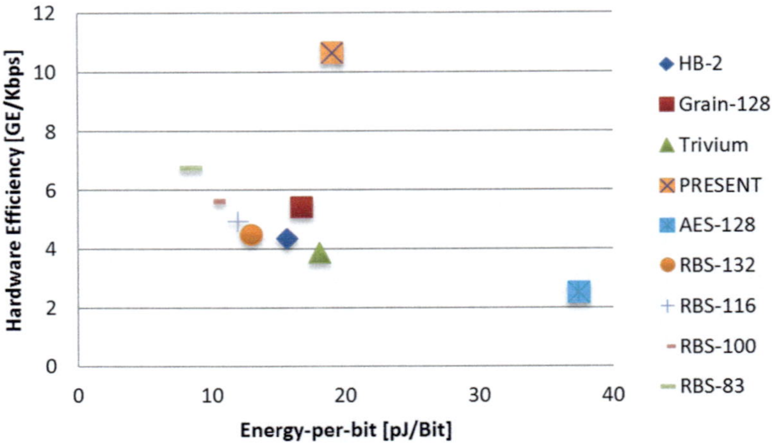

Fig. 6.14 Energy-per-bit vs. hardware efficiency without authentication

energy-per-bit while other ciphers need at least 2.5 times more energy-per-bit compared to RBS.

Comparing Figs. 6.12 and 6.13, the energy-per-bit for RBS ciphers in Fig. 6.13 is about 50 % less than in Fig. 6.12. The reason is counting out the redundant bits in the number of output bits which has direct effect on the bits-per-cycle.

6.2.8 Trade-offs

For future wireless network applications, battery life, throughput and area are three most important metrics to the designer [5]. Therefore, considering the trade-off between the energy-per-bit and throughput/area metrics may be a good measure for comparing designs.

Figure 6.14 shows the energy-per-bit metric versus the hardware efficiency of nine ciphers when the frequency is 10 MHz and all competitors do not provide the authentication service. In this comparison, the best ciphers are located in the most left and most up in the figure which need less energy-per-bit while providing higher hardware efficiency. Figure 6.14 shows that RBS ciphers have the best energy-per-bit while their hardware efficiency is medium. On the contrary, PRESENT with a big difference has the best hardware efficiency while its energy-per-bit is worse than other ciphers except AES. Among all ciphers, AES is the worst cipher in terms of both metrics.

Figure 6.15 shows the energy-per-bit metric versus hardware efficiency of four ciphers when the frequency is 10 MHz and they all provide the authentication service for 64-bit plaintexts. In this figure, RBS is the best cipher in both metrics while there is a big difference between RBS and other ciphers in both metrics.

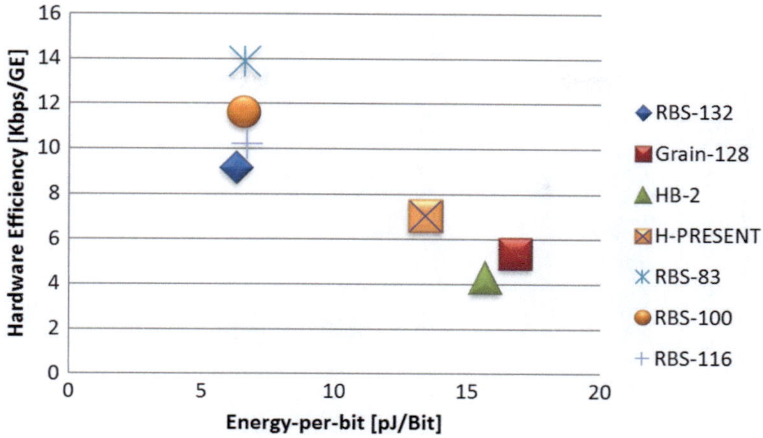

Fig. 6.15 Energy-per-bit vs. hardware efficiency with authentication

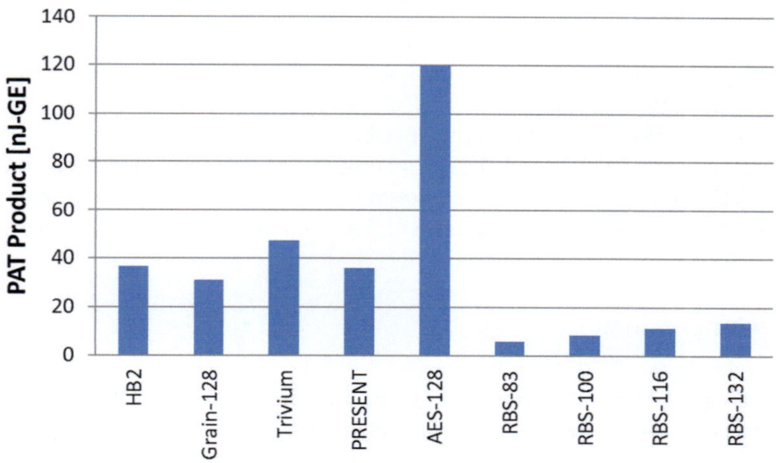

Fig. 6.16 Power-area-time product when the operating frequency is 10 MHz without authentication

6.2.9 *Power-Area-Time Product*

Power-area-time metric is a triple product calculated by multiplying the area-time product with the power consumption. It is expressed in [nJ-GE]. To have a better result in this metric, it is required that a cipher with small size in hardware consumes low power consumption with high performance [5]. In this metric, the cipher which has the lower product is the better cipher. Figure 6.16 shows the results of the power-area-time product when the operating frequency is set to 10 MHz and all competitor ciphers do not provide authentication service.

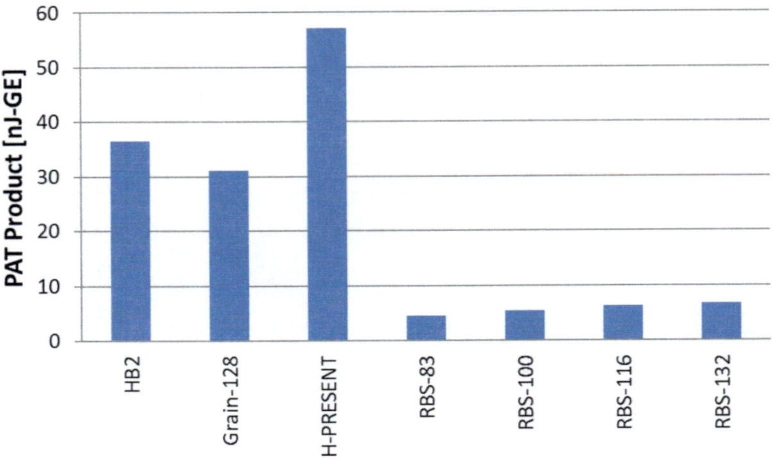

Fig. 6.17 Power-area-time product when the operating frequency is 10 MHz with authentication

In this metric, RBS ciphers have the lowest product among all competitors, while AES has the highest product.

Figure 6.17 shows the results of the power-area-time product when the operating frequency is set to 10 MHz and all ciphers provide the authentication service. In this metric, RBS ciphers have the lowest product among all other ciphers. Meanwhile, there is a huge difference between RBS ciphers and other ciphers in this metric.

6.3 Conclusions

In this chapter, the result of the hardware implementation of RBS ciphers was compared with well-known lightweight ciphers designated for RFID systems which promise high security levels and also have low cost and overhead in area and power consumption for their implementation. These ciphers had already been studied in Chap. 3 including HB-2, Grain, Trivium, PRESENT and AES. The AES cipher was considered as a benchmark cipher in our comparative study.

Since RBS is an authenticated cipher, the comparison was investigated in two categories. First, when all ciphers except RBS do no provide the authentication service. Second, when all ciphers in the comparison provide the authentication service along with confidentiality. Among these ciphers AES and Trivium do not have any dedicated hash function or authentication part and they need to be integrated with other MAC algorithms to provide this service while implementing them imposes a huge overhead in area and performance. Therefore, in the second category, these two ciphers were not included, and only four ciphers, HB-2, Grain, H-PRESENT and RBS were considered in the comparison. To have a fair competition in the second group, it was supposed that all ciphers encrypt and authenticate 64-bit plaintexts.

Table 6.17 give a summary of all the one-dimensional metrics such as the area, bit-per-cycle, throughput, maximum throughput and estimated power consumption and also the multi-dimensional metrics including the required energy for encrypting a 64-bit plaintext, the energy-per-bit, the area-time product, the hardware efficiency and the power-area-time product. In this table, the results are obtained for the 130 nm technology and 10 MHz operating frequency when ciphers do not provide the authentication service.

In Table 6.17, PRESENT is first cipher at four metrics: the bits-per-cycle and throughput, the area-time product and the hardware efficiency, while RBS is the best at the remaining metrics such as the area, the maximum throughput, the power consumption, the energy-per-bit and the power-area-time product. In conclusion, RBS ciphers have the best results in most of the performance metrics when the authentication is not required, and hence, it is strongly recommended as a candidate cipher. However, considering the energy required for transmitting the extra bits as MAC along with the ciphertext, HB-2 is the best cipher in this condition since it needs the least power consumption. On contrary, PRESENT has the best throughput among all ciphers. However, PRESENT is not a good choice compared to other ciphers because of its high power consumption and energy-per-bit.

Table 6.18 gives a summary of all one-dimensional metrics such as the area, bits-per-cycle, throughput, estimated power consumption and also multi-dimensional metrics such as the required energy for encrypting a 64-bit plaintext and providing its authentication, energy-per-bit, area-time product, hardware efficiency and power-area-time product. In this table, the results are obtained in 130 nm technology and 10 MHz operating frequency when all ciphers provide the authentication service besides confidentiality.

In Table 6.18, H-PRESENT is best at just two metrics: bits-per-cycle and throughput while RBS ciphers are the best at the rest of metrics when all ciphers provide the authentication plus confidentiality. In this table, the only metrics that RBS ciphers do have best results are bits-per-cycles and throughput. However, the result of these two metrics in RBS is very close to HB-2 and Grain.

Comparing the results in the last two tables shows that RBS ciphers are better than other ciphers especially when providing the authentication along with confidentiality is required. However, using the RBS cipher is recommended when the size of the plaintext is between 40–64 bits and the required space key is between 2^{80}–2^{128}. For environments that need very short plaintext with a large key space, RBS may not be a good choice. Because, it is required to either insert a big number of redundant bits or select a big size block of data to provide this large space key. Both solutions will impose a lot of overhead in the hardware implementation and also the transmission of the output. On contrary, stream ciphers like Grain and Trivium are very good choices when the size of plaintexts is very small (e.g., few bits) and the key size is large. However, using these ciphers without providing the authentication service is not a proper choice as if one or more bits of the ciphertext are flipped during the transmission either by accident or by an attacker, the integrity of the generated ciphertext cannot be verified at the receiver. Trivium does not have any dedicated hardware for this purpose, and hence, it is not recommended at all.

Table 6.17 Summary of normalized metrics without authentication

	Area [GE]	Bits per cycle	Max. throughput [Mbps]	Throughput [Mbps]	Estimated power [μW]	Energy/bit [pJ/bit]	Area-time product [GE-μs]	Hardware efficiency [Kbps/GE]	Power-area-time product [nJ-GE]
HB-2	2332	1	N/A	10	156.8	15.68	233.2	4.29	36.56
Grain-128	1857	1	925.9	10	167.73	16.77	185.7	5.38	31.147
Trivium	2599	1	358.4	10	181.18	18.11	259.9	3.85	47.088
PRESENT	1884	2	N/A	20	382.86	19.13	94.2	10.61	36.065
AES-128	3200	0.8	104	8	300	37.5	400	2.5	120
RBS-83	688	0.46	1150	4.6	39.34	8.55	149.6	6.69	5.88
RBS-100	826	0.46	1150	4.6	47.1	10.24	179.2	5.57	8.44
RBS-116	951	0.47	1175	4.7	56.5	12.02	203.5	4.94	11.5
RBS-132	1061	0.47	1175	4.7	60.92	12.96	225.4	4.43	13.73
Better is:	Lower	Higher	Higher	Higher	Lower	Lower	Lower	Higher	Lower

Table 6.18 Summary of normalized metrics without authentication

	Area [GE]	Bits per cycle	Throughput [Mbps]	Estimated power [μW]	Energy [pJ]	Energy/bit [pJ/bit]	Area-time product [GE-μs]	Hardware efficiency [Kbps/GE]	Power-area-time product [nJ-GE]
HB-2	2332	1	10	156.8	200.7	15.68	233.2	4.29	36.56
Grain-128	1857	1	10	167.73	214.6	16.77	185.7	5.38	31.147
H-PRESENT	4256	3	30	402.42	257.5	13.41	141.8	7.05	57.06
RBS-83	688	0.95	9.54	62.94	87.61	6.6	72.12	13.86	4.54
RBS-100	826	0.96	9.61	62.8	87.1	6.53	85.95	11.63	5.397
RBS-116	951	0.97	9.66	64.57	88.55	6.68	98.45	10.16	6.357
RBS-132	1061	0.97	9.7	60.92	82.85	6.28	109.3	9.14	6.658
Better is:	Lower	Higher	Higher	Lower	Lower	Lower	Lower	Higher	Lower

Table 6.19 The size of ciphertext for different input sizes when authentication is provided

	16 bit	32 bit	48 bit	64 bit	96 bit	128 bit
HB-2	16+64	32+64	48+64	64+64	96+64	128+64
Grain	16+32	32+32	48+48	64+64	96+96	128+128
H-PRESENT	64+128	64+128	64+128	64+128	(64+128)*2	(64+128)*2
RBS	83	83	100	132	208	272

Until now, all of the studied performance metrics were related to the hardware implementation results which are required for the computation of the ciphertext. However, extra resources are also required besides these metrics for transmitting the ciphertext which depends on the length of the ciphertext. Even though the experimental results show that RBS ciphers are the best cipher for computing the ciphertext along with authentication, it is required to compare it with other ciphers for a new metric that is the length of the output which is counted as an overhead in transmitting messages.

The length of the output in cryptosystems is equal to the length of the ciphertext plus the length of the MAC if it is provided. Adding the MAC to the ciphertext for the sake of providing authentication service will make the output longer than before. Since extra power is required for transmitting each bit of the message, the total energy required for transmitting the authenticated message will be higher than when authentication is not provided. Table 6.19 shows the length of the output for different ciphers with different input sizes. For each entry, the first number shows the length of the ciphertext and the second number shows the length of the MAC.

The relationship between the size of the plaintext and the size of the output composed of the ciphertext and MAC is shown in Fig. 6.18. Based on this figure, PRESENT cipher generates the longest authenticated ciphertexts among all ciphers. Therefore, PRESENT is not a good cipher at all for environments in which authentication is necessary because of the huge overhead in the length of MAC and also the energy required for sending it out.

For plaintexts shorter than 40 bits, Grain has the shortest output while for plaintexts longer than 64 bits HB-2 has the shortest output. For plaintexts between 40 and 64 bits HB-2, Grain and RBS ciphers produce about the same size of output. It must be noticed that for different ranges of plaintext, different RBS designs are utilized.

Based on Fig. 6.18 when authentication is required, it can be concluded that for messages shorter than 40 bits Grain is a good choice, and for messages longer than 64 bits HB-2 is a good candidate because of the short MAC, and consequently, the short output they generate. However, for messages between 40–64 bits, RBS is a good cipher because the length of its output and also the energy required for sending it out is the same as HB-2 and Grain ciphers while the energy required for computing the output is very less than its competitors.

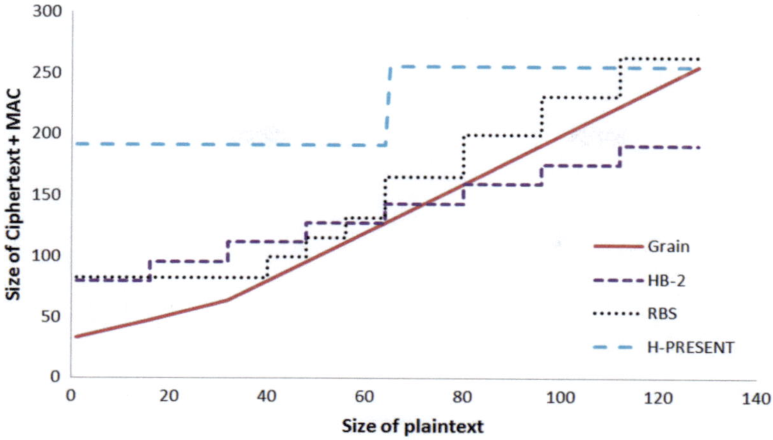

Fig. 6.18 Size of the output for different sizes of the plaintext

References

1. Bogdanov, A., Knudsen, L., Leander, G., Paar, C., Poschmann, A., Robshaw, M., Seurin, Y., Vikkelsoe, C.: PRESENT: an ultra-lightweight block cipher. In: Cryptographic Hardware and Embedded Systems - CHES 2007. Lecture Notes in Computer Science, vol. 4727, pp. 450–466. Springer, Berlin (2007)
2. Bogdanov, A., Leander, G., Paar, C., Poschmann, A., Robshaw, M., Seurin, Y.: Hash functions and RFID tags: mind the gap. In: Cryptographic Hardware and Embedded Systems - CHES 2008. Lecture Notes in Computer Science, vol. 5154, pp. 283–299. Springer, Berlin (2008)
3. Engels, D., Saarinen, M., Smith, E.: The Hummingbird-2 lightweight authenticated encryption algorithm. In: Proceedings of Workshop on RFID Security (RFIDSec) (2011)
4. Feldhofer, M., Rechberger, C.: A case against currently used hash functions in RFID protocols. In: On the Move to Meaningful Internet Systems 2006: OTM 2006 Workshops. Lecture Notes in Computer Science, vol. 4277, pp. 372–381. Springer, Berlin (2006)
5. Good, T., Benaissa, M.: Hardware results for selected stream cipher candidates. In: State of the Artof Stream Ciphers 2007 (SASC 2007), Workshop Record, pp. 191–204 (2007)
6. Hamalainen, P., Alho, T., Hannikainen, M., Hamalainen, T.D.: Design and implementation of low-area and low-power AES encryption hardware core. In: Proceedings of the 9th EUROMICRO Conference on Digital System Design, DSD '06 (2006)
7. Hell, M., Johansson, T., Meier, W.: Grain - a stream cipher for constrained environments. Int. J. Wire. Mob. Comput. **2**(1), 86–93 (2007)
8. Juels, A., Weis, S.: Authenticating pervasive devices with human protocols. In: Advances in Cryptology - CRYPTO 2005. Lecture Notes in Computer Science, vol. 3621, pp. 293–308. Springer, Berlin (2005)
9. Kaps, J.P., Sunar, B.: Energy comparison of AES and SHA-1 for ubiquitous computing. In: Emerging Directions in Embedded and Ubiquitous Computing. Lecture Notes in Computer Science, vol. 4097, pp. 372–381. Springer, Berlin (2006)
10. Satoh, A., Inoue, T.: ASIC-hardware-focused comparison for hash functions MD5, RIPEMD-160, and SHS. In: Proceedings of International Conference on Information Technology: Coding and Computing (ITCC) (2005)
11. Zhilyaev, S.: Evaluating a new MAC for current and next generation RFID. Master thesis, University of Massachusetts - Amherst (2010)

Chapter 7
RBS RFID Security and the Internet of Things

Abstract The Internet of Things (IoT) is considered as the next generation of science revolution. Conceptually, it is possible to enable Internet connectivity to anything such as cloths, TVs, machines, cars, ..., etc. IoT is a futuristic application of Wireless Sensor Network (WSN). The main characteristic of WSNs is that it is an ad-hoc network for enabling the communication of humans with the surrounding physical environment. On the other hand, Radio Frequency Identification (RFID) is considered the replacement technology for the traditional identification techniques such as barcode systems. One of the challenging research areas is the integration of both IoT and RFID technologies together such that it will be possible to have the advantages of both of worlds. This chapter discusses how does the unique characteristics of our RBS lightweight cryptosystem makes it a strong candidate of RFID security in IoT applications. The chapter then explains how to integrate IoT systems with the RFID technology and the pros and cons of such integration from security point of view.

The Internet of Things (IoT) is considered as the next generation of science revolution. Conceptually, it is possible to enable Internet connectivity to anything such as cloths, TVs, machines, cars, ..., etc. IoT is a futuristic application of Wireless Sensor Network (WSN). The main characteristic of WSNs is that it is an ad-hoc network for enabling the communication of humans with the surrounding physical environment. On the other hand, Radio Frequency Identification (RFID) is considered the replacement technology for the traditional identification techniques such as barcode systems. One of the challenging research areas is the integration of both IoT and RFID technologies together such that it will be possible to have the advantages of both of worlds. This chapter discusses how does the unique characteristics of our RBS lightweight cryptosystem makes it a strong candidate of RFID security in IoT applications. The chapter then explains how to integrate IoT systems with the RFID technology and the pros and cons of such integration from the security point of view.

A. Khattab et al., *RFID Security*, Analog Circuits and Signal Processing,
DOI 10.1007/978-3-319-47545-5_7

7.1 RBS Characterizing Features

In the previous chapters, a new lightweight symmetric authenticated encryption cipher called Redundant Bit Security (RBS) has been proposed for RFID systems. RBS provides authentication, integrity and confidentiality services at the same time with low overhead in terms of area and power consumption. The RBS algorithm presents block ciphers working with 40–64 bit data blocks and 83–132 bit keys.

In the RBS algorithm, confidentiality of the plaintext is accomplished by changing the location of plaintext bits through inserting some redundant bits inside the plaintext. The location of the redundant bits inside the ciphertext is the secret key shared between the two communicating parties. To distinguish the plaintext from the redundant bits, the attacker will need accessing this key. Obviously, increasing either the size of the plaintext or the number of redundant bits, the number of possible keys for the attackers will increase. Therefore, these two parameters have important effects on the security strength of RBS. In other words, the security level of the RBS algorithm is adjustable through these parameters.

Besides confidentiality, the redundant bits provide other services as well such as integrity and authentication. Hence, the receiver can validate the data if the data gets altered during transmission either by an adversary or by other environmental issues. To fulfill all these services, the redundant bits are defined as the message authentication code (MAC) of the plaintext.

The MAC algorithm in RBS is based on the proposed MAC in [1] as it supports different digest sizes. Supporting different digest sizes means different redundant data sizes, and accordingly, different security levels could be adjusted without changing the underlying MAC algorithm. However, there is a limitation using the MAC algorithm in [1] as it is mainly proposed for stream cipher systems. In order to adapt it with RBS block cipher algorithm, some modifications has been performed on this authentication approach. Rather than adapting the algorithm, these modifications saved considerable resources in the hardware in terms of area and power at the cost of some marginal throughput performance.

Besides the redundant data, how the plaintext appears inside the ciphertext also contributes in the confidentiality of RBS. In RBS, the plaintext gets altered through bitwise addition of the plaintext bits with a variable keystream which itself is a function of the plaintext before merging with the redundant data. This way, any change in the plaintext results in a different keystream, and consequently, a different ciphertext. The altered plaintext is also generated through the MAC as the case with the redundant data.

The security of RBS against powerful and well known attacks such as known-plaintext attack, chosen-plaintext attack, chosen-ciphertext attack, differential attack, substitution attack, related key attack, linear cryptanalysis algebraic attack and cube attack has been proved in this book.

The main hardware component contributing in resource usage in RBS is the MAC generator which is responsible to produce the redundant bits and the keystream. Rather than the MAC hardware block, there are encryption/decryption components for inserting/separating the redundant bits to/from the altered plaintext which are embedded in the sender/receiver. These two parts are composed of a multiplexer and de-multiplexer which can be implemented by few gates.

Compared to other existing symmetric ciphers for RFID systems, RBS offers the lowest area cost by decreasing the area to 43 % which is concluded from serialization input, resource sharing and using simple elements such as XOR and multiplexers. Reducing the cost of area results in 53 % power saving with only 3 % performance degradation. When authentication is not required, RBS is better than other ciphers in terms of area, power, energy, area-time product and power-area-time product and the worst in performance and throughput. Also, the cost of energy for transmission is about two times more than other ciphers. When authentication is a must like hostile environments or payment systems, the performance and throughput of RBS is very close to other ciphers. These superior features of RBS makes it a strong candidate for integration in many futuristic IoT applications especially those in need of low-power and lightweight cryptosystem.

7.2 RBS Future Extensions

Throughout this research, some ideas may expand the scope of original goals and mitigate some restrictions of RBS. This section provides an overview of possible ideas that could be followed in further work.

- One of the limitations in RBS is the length of the generated ciphertext. Making a hybrid version of RBS cipher, the value of the MAC will depend on the initial vector such that there would be no worry about MAC collisions. In other words, the size of the MAC can be the same as the plaintext size like Grain cipher. However, the main challenge would be the resilience of the algorithm against chosen-plaintext attacks.
- The underling serialized architecture of RBS is the main source of its low throughput. Instead of processing one input bit at each clock cycle, two or more bits can be processed by adding some parallel hardware resources to the system. It is predicted that by increasing the cost of area by 30–40 %, the performance can be approximately doubled.
- In RBS, the location of the plaintexts changes by inserting the redundant bits inside the ciphertext while their order is intact. The security of RBS cipher can be improved considerably by changing the order of plaintext bits as it will increase the key space size exponentially. The main challenge for this approach would be defining keys that store the information of bits order as well as their location in the ciphertext.
- In the presented algorithm, the same key is used for generating authentication, keystream and encryption processes. Splitting the key into sub-keys will increase the key space. However, these keys are required to be divided and utilized somehow that any change at each of the sub-keys has significant change in both the generated redundant data and the keystream in order to be strong against related key attacks.

7.3 The Internet of Things (IoT)

The Internet of Things (IoT) is the concept of physical objects or "things" equipped with electronics, software and sensors with network connectivity as shown in Fig. 7.1. Each "thing" is uniquely distinguishable through its embedded computing system but is able to interoperate within the existing Internet infrastructure. This enables such objects to be sensed and controlled remotely across the existing network infrastructure, creating opportunities for more direct integration between the physical world and computer-based devices.

Connecting the different "things" to the Internet to increases our control on these things, changing them to be smart and intelligent, provides new areas of development by proposing new applications for approaching the maximum automation and control. It enables the exchange of data never available before, and brings users' distributed information to a more centralized platform, that can lead to the introduction of the concept of big data to analyze these huge amounts of data getting new layer of valuable statistics. It is estimated that the IoT will reach 50 billion devices connected to the Internet by 2020. Furthermore, gaining deeper insight with analytics of the collected IoT data enhances human race productivity, creates new business models, and generates new revenue streams.

A typical IoT system architecture is composed of the following main components illustrated in Fig. 7.2:

- **Objects or Things:** The objects or the things are mainly whatever devices, appliances, pets, . . . , etc. that will be carrying a communication module in order

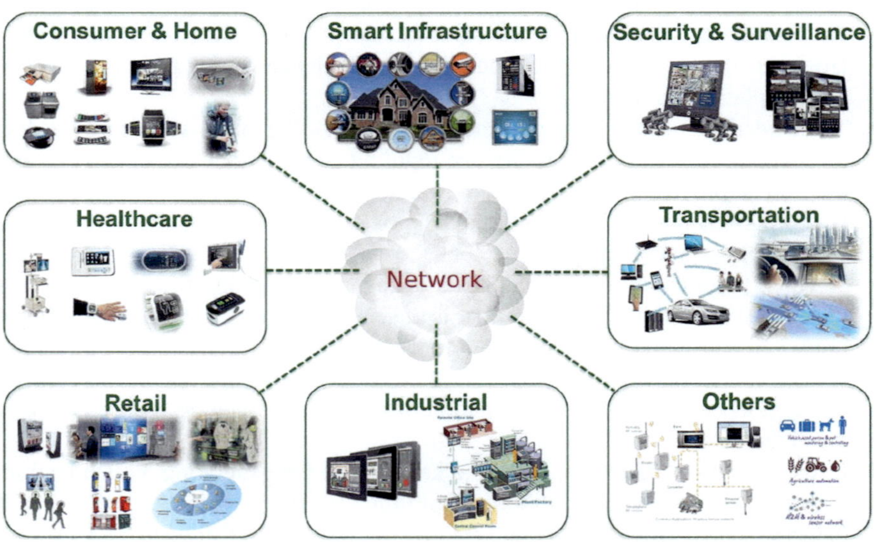

Fig. 7.1 The Internet of Things (IoT) paradigm

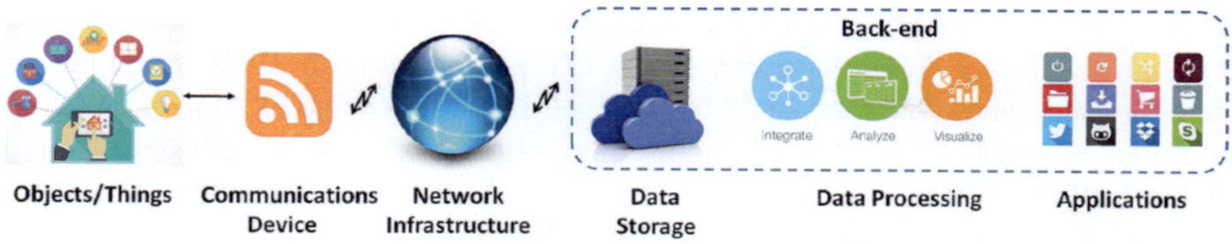

Fig. 7.2 Typical IoT system architecture

to make them smarter. The objects/things are the elements that will be connected to the Internet to provide with more value and functionality.

- **Communication Module:** This is the hardware or the end terminal that allows the objects/things to communicate with the Internet infrastructure. It allows identifying the addressed object and accessing a particulary required service on the object to allow a range of features that can be provided, hence, it must be intelligent. Communication modules can be fabricated by any company taking in consideration the communication protocol standards and the ability to send the required output.
- **Network:** A network refers the communication infrastructure that provides connectivity of the things to the Internet for reaching the end servers to access their services.
- **Back-end** In order to establish an end-to-end service between the object and other interested communicating party, it is important to have a back platform that is responsible for the following tasks:
  - *Data Storage:* It is of crucial importance to store the huge amount of received information sent from the objects/things via the communication device in order to start processing such data and develop innovative applications.
  - *Data Processing:* The most valuable step of an IoT system is that the huge amount of collected data to be processed to start getting more valued information. This step allows several businesses to emerge to provide the customers with several application based on the processed information.
  - *Applications:* The application servers are needed to provide the new or valuable services to the IoT system users to interact with their objects/things through the network connectivity provided by any of the wireless/wired Internet service providers.

7.3.1 IoT History

Over the last few years, the Internet of Things has evolved from an intriguing concept into an increasingly sophisticated network of devices and machines as more and more "things" get connected to the Internet. The Internet of Things has evolved through years of developing plans. In 1990, John Romkey and Simon Hackett created the world's first connected device (other than a computer): a toaster powered through the Internet. The concept of the IoT gets a name in 1999, when Kevin Ashton establishes Massachusetts Institute of Technology (MIT) Auto-ID Center, a global research network of academic laboratories focused on Radio Frequency Identification (RFID).

Also, in 1999 the machine-to-machine (M2M) protocol was developed. Andy Stanford-Clark of IBM and Arlen Nipper of Arcom (now Eurotech) introduced the first machine-to-machine protocol for connected devices: MQ Telemetry Transport (MQTT). In 2000, LG introduced a big chill when announced plans for the first connected refrigerator. The fridge will sense items stored inside it using barcode and RFID scanning.

In 2005, the United Nations (UN) mentioned the IoT where it got global attention in an International Telecommunications Union (ITU) report. Three years later, the first international IoT conference takes place in Zurich.

In 2008, The IPSO Alliance (non-profit organization founded with members from technology, communications and energy companies) was formed to promote IP connections across networks of "smart objects". The alliance now boasts more than 50 member firms.

In 2010, Google introduced a self-driving vehicle project that was a major milestone in the development of a connected and autonomous car. Also in 2010, Bluetooth Low Energy (BLE) is introduced, enabling applications in the fitness, health care, security, and home entertainment industries.

In 2011, Nest Labs (now Google) introduced sensor-driven, WiFi-enabled, self-learning, programmable thermostats and smoke detectors. Also in 2011, IPv6 was launched, The protocol expands the number of objects that can connect to the Internet by introducing 340 undecillion IP addresses (2^{128}).

In 2013, Google Glass (an optical head-mounted display that is worn like a pair of eyeglasses), controlled through voice recognition software and a touchpad built into the device is released to developers.

In 2014, Apple announces Health Kit and Home Kit for health and home automation developments. The firm's iBeacon advances context and geolocation services.

Table 7.1 illustrate the evolution of the number of devices connected to the Internet over the years.

Table 7.1 History of IoT connectivity

Year	World population [Billion]	Connected devices [Billion]	Connected devices/person
2003	6.3	0.5	0.08
2010	6.8	12.5	1.84
2015	7.2	25	3.47
2020	7.6	50	6.58

7.3.2 *IoT Challenges*

Despite the promises of the IoT technology, it is faced by many challenges including:

- **IoT Devices Cost:** For the wide adoption of IoT services, the end devices must be affordable by wide segments of the population, and also should consume small amounts of power to be able to operate for a long time while minimizing the operational expenditure (OPEX). Another factor is the reliability of the IoT devices.
- **Addressing:** One might question the ability to address all this expected huge number of devices to the Internet with the limited address space of IPv4 (which allows for 4.3 billion unique addresses). In fact this implies that objects in the IoT will have to use IPv6 to accommodate the extremely large address space required. However, IPv6 is not suitable for the resource-limited IoT devices.
- **Global Standards:** Since heterogenous devices are planned to communicate with each other, many protocols need to be set while regulating their interactions. This adds to the current developed protocols for IoT objects interaction with the Internet. Two examples are:
  - *REpresentational State Transfer (REST)* is a scalable architecture which allows for things to communicate over Hypertext Transfer Protocol (http) and is easily embraced in IoT applications to provide communication from a "thing" to a central web server.
  - *Message Queuing Telemetry Transport (MQTT)* is an architecture on top of TCP/IP which allows for bi-directional communication between a "thing" and a MQTT broker which is responsible for distributing messages to interested clients based on the topic of the message.
- **Security:** When the world's billions of sensors are constantly acquiring data from their surroundings, which includes humans, then privacy concerns are paramount in an IoT world. Most of the developed world has attempted to protect consumers from illegal use of confidential information, but in many cases the laws are not adequate to meet the tremendous number of new ways personal information is being captured and used. Concerns have been raised that the IoT is being developed rapidly without appropriate consideration of the intense security challenges involved. In particular, cyber-attacks are likely to become a physical (rather than simple virtual) threat. Many Internet-connected appliances can already "spy on people in their own homes" including televisions, kitchen appliances and cameras through network attackers.

7.3.3 *Applications*

The Internet of Things have applications in almost every aspect of man's modern life [2, 5]. The following are just a few of such IoT applications

- **Energy Consumption Optimization:** Energy consumption control is one of the most important benefits of the IoT. It is used in different applications to save the energy in any environment by putting smart sensors and machines to provide services that will insure energy conservation that is estimated to be near 40 %. This application basically refers to accessing information about energy consumption and reacting to the information to optimize the allocation of energy use. Once the residents know they have been using their washing machine during peak hours when the grid is most constrained and the cost of electricity is at premium, they could adjust their behavior and wash their laundry during non-peak hours, saving money and helping the utility company cope with the peak demand.
- **Health care:** For health care IoT applications, new smart intelligent machines that provide new functions that will improve the application supporting the customer's health and assisted living. New watches or bracelets are used to measure the body heart rates, blood pressure, blood sugar and others measurement getting reports to be sent to the doctors saving time, money and effort trying to catch them up and provide a better care for the health of the customer. Monitoring devices for various patients and different body parts are all reported online or through mobile applications. They can even take a decision like sending an urgent short message (SMS) to certain doctors' contacts list if a specific key counter; e.g., blood pressure, has reached a predefined threshold.
- **Smart Meters:** This application could be assumed to be the most implemented application of all IoT applications. Power, water or gas meters are now reporting all readings to the central units online with the ability to recharge users credit or disable the service through remote controlling systems.
- **Fleet Management:** Vehicles are equipped with electronic chips to report in real-time the fleet destinations, routes and utilization.
- **Security and Surveillance:** Cameras are now able to be monitored online with live shots from the surroundings, giving advanced controls remotely to adjust azimuth or cameras orientation.
- **Agriculture:** The IoT can be used in agriculture fields, with the goal of optimizing production and efficiency while reducing costs and environmental impacts. For farming operations, it involves the analysis of real-time data on weather, soil and air quality.

7.4 RFID Systems in Internet of Things (IoT)

The Internet of Things objective is to interconnect all physical objects together (e.g., humans, refrigerators, vehicles, air-conditioners, lamps, ..., etc.) [3]. It is obvious that the RFID systems will play an important role in IoT. RFID systems

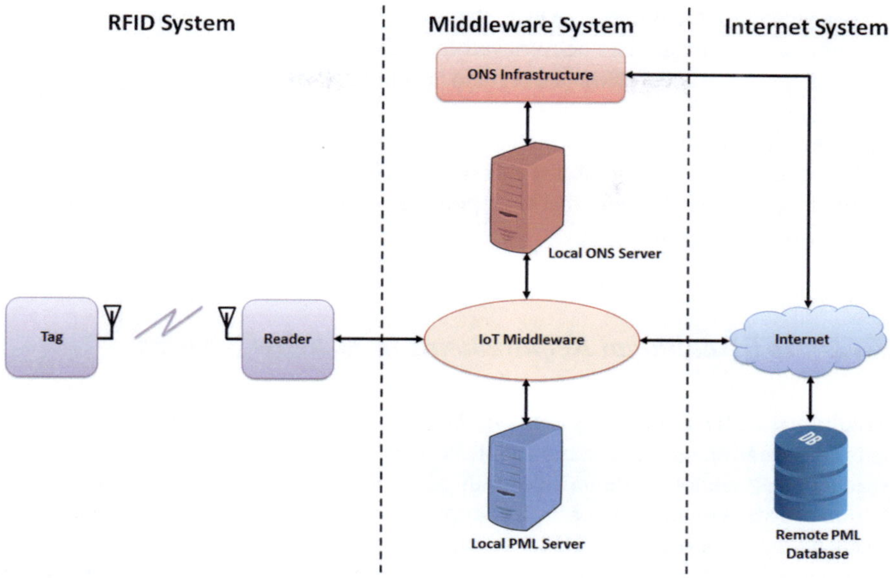

Fig. 7.3 RFID IoT-based system architecture

in IoT will open the door for new applications which will make use of the RFID systems identification capability and the IoT capabilities. On the other hand, these applications will need some additional requirements from RFID systems. In this section, we will describe the architecture of IoT based on RFID systems, will show the extra-things required from the RFID systems to enter the IoT applications world, and will highlight the security of RFID systems in IoT.

7.4.1 The Architecture of IoT Based on RFID

As shown in Fig. 7.3, the architecture of IoT based RFID systems consists of three main components [7]:

- **RFID Systems:** which are responsible for object identification and environment monitoring (optional, if the sensing capability is allowed). These consist of Electronic Product Code (EPC) based tags and readers as discussed in more details in Chap. 1.
- **Middleware System:** that is responsible for protocol switching and acts as a connecting link between the RFID systems and the Internet system. The middleware system basically consists of an Object Naming Service (ONS) and a Physical Markup Language (PML) service. The middleware system converts the EPC code to a corresponding IP-address using the ONS infrastructure

connected to the Internet. Once the EPC is converted to an IP-address any relevant information can be retrieved from the object using this address. A local ONS server, a local PML server and a remote PML server are used for data storage.

- **Internet System:** through which end users and devices can connect to the objects to retrieve their data or alter their content. This is achieved by passing requests to the middleware system that is responsible for protocol conversion and address information resolution.

7.4.2 IoT Additional Requirements from RFID Systems

Traditional RFID systems act as an *one-hop, centralized network* wherein the reader acts as a centralized base station and the tags could only communicate with the reader. Tags cannot communicate with each other. Likewise, the readers cannot communicate with each other. In IoT applications, RFID-enabled things could talk to each other. Consequently, RFID security in IoT needs to support extra properties such as mutual authentication, key establishment and data confidentiality. Here come the importance of our proposed RBS cryptosystem that is capable of achieving all these security services, despite being a lightweight cipher.

7.4.3 Security Issues with RFID-Based IoT Architectures

This section summarizes, the main security issues with RFID-based IoT architectures.

- **Security Issues with RFID Systems:** In the first part of this book, we have discussed the different security issues with RFID system in details. Basically, the RFID tag can leak its content to unauthorized readers, and hence, encryption is needed to ensure confidentiality. Furthermore, RFID readers should make sure that the information is transported from the right tag (not a copy), which needs mutual authentication between the reader and the tag. Also, the tag data is vulnerable to being tampered during communication, and hence, data integrity is needed. Finally, RFID systems are to be resilient to denial of service attack (DoS). Our RBS cryptosytem presented in Part II of the book provides a lightweight solution of such RFID security issues.
- **Security Issues with the Middleware System:** The most important security concern in the middleware system is to allow the RFID readers to *securely* communicate with the different middleware database servers. Furthermore, database servers need protection against DoS attacks, which is a traditional security issue.

- **Security Issues with the Internet System:** The Internet system in the RFID-based IoT architecture depicted in Fig. 7.3 is vulnerable to the legacy network security threats. Finding solution for such common threats, e.g., how to protect databases from being destroyed and how to ensure the confidentiality and the integrity of the data communicated over the Internet data, are open security problems that still needs to be addressed.

7.5 Integrating RFID in IoT Applications

Integrating RFID systems and WSNs widely used to develop IoT applications will extend the functionality of RFID by adding sensing and multi-hop capabilities. This will lead to increasing the scalability and the portability of readers and tags. On the other hand, RFID will extend the functionality of the WSN by adding a capability of detecting and identifying un-sensible objects. Moreover, RFID will add an alternative solution for WSN in harsh environments and some applications [6].

The integration of WSN and RFID can dramatically reduce the cost and the power consumption of the overall system. This integration will increase their effectiveness and will give new prospective applications. The integration of WSN and RFID can be according to the following architectures: integrating RFID tags with sensors (limited communication capability), integrating RFID tags with wireless sensor node (extended communication capability), integrating RFID readers with wireless sensor nodes, and mixing both systems.

7.5.1 RFID with Sensing Capabilities

As shown in Fig. 7.4, normal RFID tags are used with a unique identification and equipped with sensors to sense the environment in the case of integrating RFID tags with sensors architecture. This architecture has a limited communication capability which is restricted by the RFID transceiver. The used tags can be active tags (i.e., Sensor-Embedded Radio Frequency Identification (SE-RFID) which was proposed by Deng et al. [4]), passive tags (i.e., battery-less Wireless Identification and Sensing Platform (WISP) which was proposed by Sample et al. [9]), or semi-passive tags (e.g., the VarioSens that was the first semi-active RFID sensor-tag with an integrated sensor which was proposed by the German firm KSW-Microtec).

7.5.2 Integrating RFID in Sensor Node Architectures

As shown in Fig. 7.5, integrating RFID tags with wireless sensor node architecture implies that a wireless sensor node is equipped with an RFID tag. In other words, the tags can communicate with each other using wireless sensor node's transceivers

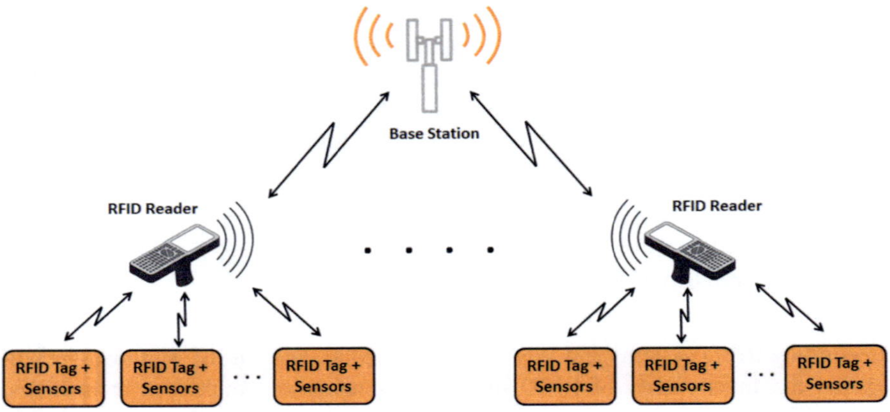

Fig. 7.4 Integrating RFID tags with sensors

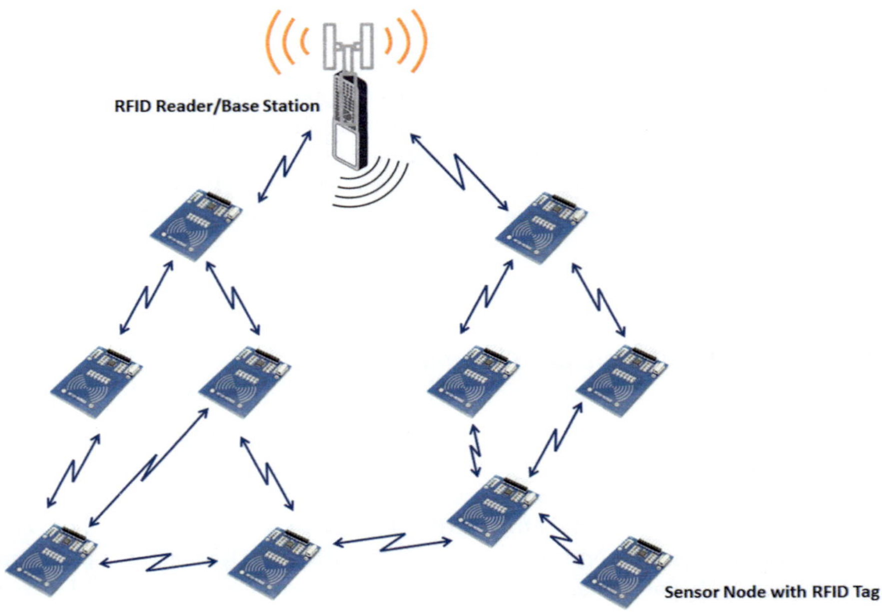

Fig. 7.5 Integrating RFID tags with wireless sensor node

(and create a cooperative ad-hoc network) as well as with RFID readers using RFID tags. This architecture has an extended communication capability and the RFID tags can communicate with each other. Ruzzelli et al. claimed that the main advantage of this architecture is adding an on-demand wakeup capability for the wireless sensor node to reduce the power consumption and to eliminate the idle listening time in

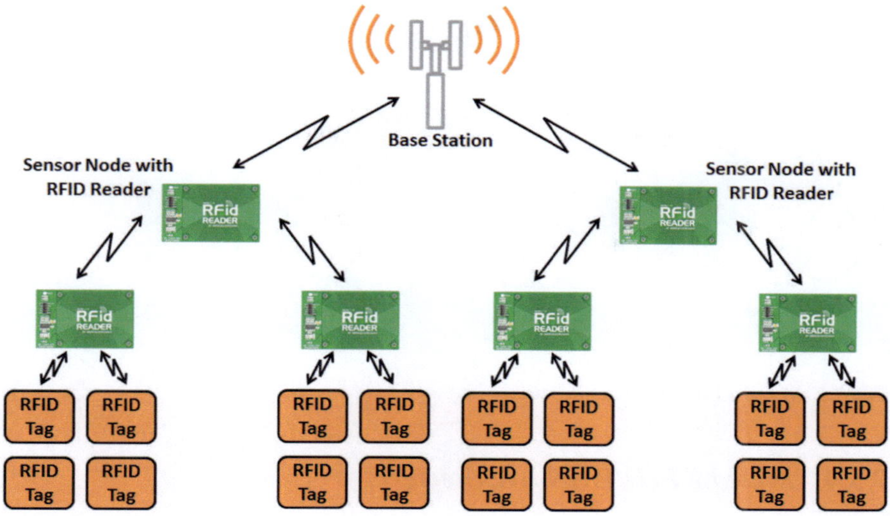

Fig. 7.6 Integrating RFID readers with wireless sensor nodes

WSNs [8]. To achieve this capability, each node has to be provided with RFID tags with reader capability (i.e. iRFID, Intelligent Radio Frequency Identification, which was proposed by Machine Talker).

7.5.3 Integrating RFID Readers in Sensor Node Architectures

As shown in Fig. 7.6, a wireless sensor node is integrated with RFID reader in the case of this architecture. There are three types of devices in this architecture: the wireless sensor/RFID reader nodes also called the smart nodes such as the SkyeRead M1-mini which proposed by SkyeTek, simple RFID tags, and the sink node. In this architecture, each smart node is a wireless sensor node which sensing capability is extended by the RFID reader. The smart nodes relay the messages read by the RFID reader using their own transceivers to reach the correct destination (mostly, the sink node) to process them. The smart nodes are less expensive, smaller, and more portable than the traditional RFID readers. This architecture will open a door for new applications. On the other hand, because of the many-to-one architecture, there will be unbalance in power consumption among the nodes. The nodes closer to the sink node will consume much more power than the remote ones. A solution of this disadvantage is to add more smart nodes close to the sink node. However, this will lead to more collisions.

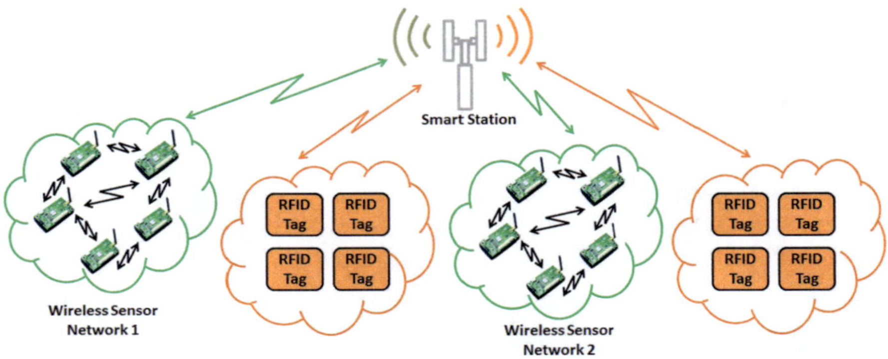

Fig. 7.7 Mixed RFID/WSN architecture

7.5.4 Mixed RFID/WSN Architecture

In mixed RFID/WSN architectures, the WSN nodes and the RFID tags are separable physically which means that there is no need to design new integrated nodes as shown in Fig. 7.7. However, they are integrated in the software layer. The advantage of this architecture is that each system will complete its function by the assistance of the other. On the other side, the physical separation will cause communication interference.

7.6 RFID-Based IoT Applications

As discussed earlier, the IoT has numerous applications. Out of these numerous applications, the following are sample applications that will benefit from the integration of the RFID technology and IoT [6]:

7.6.1 Health Care Applications

RFID and IoT can be used in health care and assisted living for the elderly applications. RFID readers can be used to locate medicine location in the home. Each medicine bottle will contain an RFID tag and a weight scale sensor to locate and detect if the bottle is empty or not. On one hand, the system can order the medicine once detecting its shortage. On the other hand, the system can send a notification or alarm to the hospital in case the elder person miss his/her dosage.

Another application related to the elder health care is a smart home prototype where there is a reader in each daily used object in the home and a tag assigned to the elder person. The tags can measure the blood pressure, temperature, glucose level, etc. The readers will detect the elder person's daily activity. In case of no activity detected or abnormal case of the elder person, the system will send an alarm to the hospital.

7.6.2 Supply Chain Applications

Sensors (temperature, humidity, etc.) can be attached to RFID tags to detect the conditions of supplies (i.e., food, medicines, etc.). In each truck there is a reader to collect the tags information and send it to a remote server to analyze it. The customer can get real time information about the supplies and get notifications/alarms in case of abnormal situations.

7.6.3 Battlefield Applications

RFID tags with sensors can be attached to each weapon to count the number of fires, their time stamp, and the heat of the weapon. A reader can get the tags' information and send it to the army servers to analyze this data and to determine whether the lifetime of the weapon is coming to an end or not.

References

1. Agren, M., Hell, M., Johansson, T.: On hardware-oriented message authentication with applications towards RFID. In: Proceedings of International Workshop on Lightweight Security & Privacy (LightSec) (2011)
2. Cisco: http://www.cisco.com (2016)
3. Das, M.L.: Strong security and privacy of RFID system for internet of things infrastructure. In: Security, Privacy, and Applied Cryptography Engineering, pp. 56–69. Springer, Berlin (2013)
4. Deng, H., Varanasi, M., Swigger, K., Garcia, O., Ogan, R., Kougianos, E.: Design of sensor-embedded radio frequency identification (SE-RFID) systems. In: Proceedings of International Conference on Mechatronics and Automation (2006)
5. IoT Council: http://www.theinternetofthings.eu/ (2016)
6. Mitrokotsa, A., Douligeris, C.: Integrated RFID and sensor networks: architectures and applications. RFID and sensor networks: architectures, protocols, security and integrations, pp. 511–535 (2009)
7. Nie, X., Zhong, X.: Security in the internet of things based on RFID: issues and current countermeasures. In: Proceedings of 2nd International Conference on Computer Science and Electronics Engineering (2013)

8. Ruzzelli, A.G., Jurdak, R., O'Hare, G.M.P.: On the RFID wake-up impulse for multi-hop sensor networks. In: Proceedings of 1st ACM Workshop on Convergence of RFID and Wireless Sensor Networks and their Applications (SenseID) at the 5th ACM International Conference on Embedded Networked Sensor Systems [ACM SenSys 2007] (2007)
9. Sample, A.P., Yeager, D.J., Powledge, P.S., Smith, J.R.: Design of a passively-powered, programmable sensing platform for UHF RFID systems. In: Proceedings of IEEE International Conference on RFID, pp. 149–156 (2007)

Glossary

Active Tag An RFID tag with a radio signal transceiver embedded along with a power source, usually in the form of a small battery to power it. An active RFID tag can initiate communication and activate itself regardless of the presence of a reader in their vicinity.

Authentication A process through which an object proves its claimed identity to other communicating parties by providing some evidence such as what it knows, what it has, or what it is.

Automatic Identification or Auto-ID A broad term that refers to any technology that can identify and locate physical objects automatically by electronically exchanging data and without any human interaction.

Block Cipher An encryption function that works on fixed size blocks.

Blocker Tag A blocker tag is a physical solution for protecting privacy in RFID systems. A blocker tag is similar to an RFID tag except that it can block readers from reading the identification of those tags that exist in the blocker tag's range.

Cryptography A fundamental method for keeping the communication between two parties private in the presence of third parties. The word "cryptography" is derived from the Greek roots, "kryptos" and "graphein" meaning secret writing.

Faraday Cage A Faraday cage is an enclosure design made of conducting materials to exclude electromagnetic fields.

Key Space The set of all possible keys that can be used to initialize a cryptographic algorithm.

Passive Tag An RFID tag that has no internal power source. It draws its power from the electromagnetic field generated by the RFID reader. It also has no active transmitter and rely only on the power that comes from a reader's signal.

Physical RFID Threats Physical threats are those threats that use physical means to attack the RFID system to disable tags, modify their content, or imitate them.

Privacy Violations Security threats in which the attacker tries to harvest information from the objects by eavesdropping to the communications between the object and the reader or by tracking them.

A. Khattab et al., *RFID Security*, Analog Circuits and Signal Processing,
DOI 10.1007/978-3-319-47545-5

Redundant Bit Security (RBS) A lightweight symmetric encryption algorithm that targets the resource-constrained RFID devices.

RFID Channel Threats Channel threats refer to the attacks targeting the insecure channel between a reader and a tag.

RFID Transponder or RFID Tag A data carrying device that is attached or embedded to the items to be later interrogated by an RFID reader.

RFID Transceiver or RFID Reader A two-way radio transmitter-receiver that both transmits and receives radio waves unlike RFID tags which transmit signals only in response to received signals.

Security Violations Security threats in which an adversary counterfeits the behavior of a tag or a reader for making undesirable communications. They may target the physical tag, the communication channel between the tag and the reader, or the application or the system which employs the RFID technology.

Semi-passive Tag An RFID tag that has its own power supply that supports the integrated microchip only. When the battery is discharged, such a tag cannot transmit signals any more. Unlike active tags, a semi-passive tags has no active transmitter.

Stream Cipher An encryption function that processes the message bit by bit as a stream.

System Threats System threats mainly refer to the attacks on the flaws existed in the authentication protocol and encryption algorithm.

About the Authors

Ahmed Khattab received his B.Sc. (honors) and M.Sc. in Electronics and Communications Engineering from Cairo University, Egypt, in 2002 and 2004, respectively. He received the Master of Electrical Engineering degree from Rice University, and his Ph.D. in Computer Engineering from the University of Louisiana at Lafayette, USA, in 2009 and 2011, respectively. He is currently an Assistant Professor at the Electronics and Electrical Communications Engineering Department in Cairo University. He is also adjunct Assistant Professor in the American University in Cairo (AUC). His research interests are in the broad areas of wireless networking with emphasis on the cross-layer design, optimization, and implementation of PHY/MAC protocols for high performance wireless networks. His research experience ranges from wireless sensor networks and the Internet of Things (IoT) to distributed opportunistic spectrum management for cognitive radio networks, carrier-sense multiple access for multi-antenna 802.11 networks, resource management and scheduling in 4G/5G wireless networks, and vehicular networks. He has authored/co-authored two books, a patent application and over 40 journal and conference publications. He serves as a reviewer in many IEEE transactions, journals and conferences, and is a member of the technical committee of several

A. Khattab et al., *RFID Security*, Analog Circuits and Signal Processing,
DOI 10.1007/978-3-319-47545-5

prestigious conferences such as IEEE Globecom, IEEE ICCCN, and IEEE WF-IoT. He won the best student paper award from the IEEE Computer Society at the University of Louisiana at Lafayette chapter in 2010 and in 2011, and was a finalist in the best paper award contest in the IEEE ICCCN 2008 conference.

Zahra Jeddi received her BS degree in Electrical Engineering from Iran University of Science and Technology and her MS degree in Computer Engineering from Amirkabir University of Technology. Zahra Jeddi joined the Center for Advanced Computer Studies (CACS) at the University of Louisiana at Lafayette where she obtained her Doctor of Philosophy degree in Computer Engineering in Summer 2014. She contributed in several research projects at CACS such as body area network, RFID security, and crypto architecture design. Her research interests include hardware security, low power design, and computer architecture. She is currently with Intel Corporation.

Esmaeil Amini received his Bachelor of Science in Computer Engineering from Sharif University of Technology, Tehran, Iran. He received a Master of Science in Computer Engineering from Amirkabir University of Technology, Tehran, Iran.

Esmaeil Amini was one of the core developers of Persia; the synthesis tool for asynchronous and Globally Asynchronous Locally Synchronous (GALS) circuits.

Esmaeil Amini joined to the Center for Advanced Computer Studies (CACS) at the University of Louisiana at Lafayette in Spring 2009. He contributed in several research projects at CACS such as body area networks, RFID security, and crypto architecture design. He completed the requirements for his Doctor of Philosophy in Computer Engineering in Spring 2013. He is currently with Yahoo corporation.

Magdy A. Bayoumi received the B.Sc. and M.Sc. degrees in electrical engineering from Cairo University, Egypt, the M.Sc. degree in computer engineering from Washington University, St. Louis, and the Ph.D. degree in electrical engineering from the University of Windsor, Ontario. He is the director of the Center for Advanced Computer Studies (CACS) and was the department head of the Computer Science Department, University of Louisiana, Lafayette. He is also the Z.L. Loflin Eminent Scholar Endowed Chair at the Center for Advanced Computer Studies, University of Louisiana, Lafayette, where he has been a faculty member since 1985. He is a fellow of the IEEE. He was the Vice President for Conferences of the IEEE Circuits and Systems (CAS) Society. He was the vice president for the technical activities of the IEEE Circuits and Systems Society and the chairman of the Technical Committee on Circuits and Systems for Communication and the Technical Committee on Signal Processing Design and Implementation. He was a founding member of the VLSI Systems and Applications Technical Committee and was its chairman. He is a member of the Neural Network and the Multimedia Technology Technical Committees. He was an associate editor of Circuits and Devices Magazine, the IEEE Transactions on Very Large Scale Integration (VLSI) Systems, the IEEE Transactions on Neural Networks, , the IEEE Transactions on Image Processing, and the IEEE Transaction on Circuit and Systems II: Analog and Digital Signal Processing and Integration. Dr. Bayoumi served on the Distinguished Visitors Program for the IEEE Computer Society, 1991–1994 and the Circuits and Systems, 1999–2001. Dr. Bayoumi is the recipient of the 2009 IEEE Circuits and Systems Meritorious Service Award and the IEEE Circuits and Systems

Society 2003 Education Award. He won the Researcher of the year award and the Distinguished Professor award from the University of Louisiana at Lafayette in 1988 and 1993, respectively. Dr. Bayoumi served on the technology panel and advisory board of the US Department of Education project, “Special Education Beyond 2010,” 1990–1993. He was the vice-president of the Acadiana Technology Council. He was on the governor’s commission for developing a comprehensive energy policy for the State of Louisiana. He represented the CAS Society on the IEEE National Committee on Engineering R&D policy, IEEE National Committee on Communication and Information Policy, and IEEE National Committee on Energy Policy. Dr. Bayoumi research interests include VLSI design and architectures, low power circuits and systems, digital signal processing, neural networks, and wireless ad-hoc and sensor networks.

Index

A. Khattab et al., *RFID Security*, Analog Circuits and Signal Processing,
DOI 10.1007/978-3-319-47545-5

O

P

R

S

T

W

MIX
Papier aus verantwortungsvollen Quellen
Paper from responsible sources
FSC® C105338

If you have any concerns about our products,
you can contact us on
ProductSafety@springernature.com

In case Publisher is established outside the EU,
the EU authorized representative is:
Springer Nature Customer Service Center GmbH
Europaplatz 3, 69115 Heidelberg, Germany

Printed by Libri Plureos GmbH
in Hamburg, Germany